Life Sciences and Health Challenges

New York Academy
of Sciences

Life Sciences and Health Challenges

Edited by
SUSAN RAYMOND, PH.D.

New York Academy
of Sciences

New York Academy of Sciences
Two East Sixty-third Street
New York, New York 10021
telephone: 212.838.0230, ext. 348
fax: 212.753.3479; e-mail: policy@nyas.org

ISBN 1-57331-148-0

Contents

Life Sciences and Health Challenges

Preface

I n 1787, Thomas Jefferson wrote, "With your talents and industry, with science, and that steadfast honesty which pursues right, regardless of consequences, you may promise yourself every thing—but health, without which there is no happiness." It is understandable, then, that health—and the professions, organizations, and budgets that serve and surround it—should be a central concern of nations and of the world. Health care is in transition. Demographics are changing. The world is aging. Resources are strained. Organizational methods for preventing and curing disease (and then paying the bills) are constantly shifting.

Throughout its more than 180 years, the New York Academy of Sciences has advanced the evolution of biomedical sciences and the application of innovation to provision of health care. The Academy's roots in the research are deep, and its commitment to building bridges between life sciences, societal and economic change, and health policy is lasting.

Over many years, a variety of national and international leaders—public, corporate, and academic—have come to the Academy to discuss the future of national and global health, and the challenges likely to be faced by science, technology, and public policy in navigating into that future. The essays in this special volume reflect recent discussions, as well as original analytic work by the Academy itself.

In many quarters and for most people, health care is like the weather. Everyone complains, but no one does any thing about it. For the Academy, and for the leaders who participate in its policy forums, that characterization is unacceptable. This volume presents not only thoughtful insights on health conditions, it also offers pragmatic suggestions for improvements to the local, national, and global organizations charged with providing, managing, and paying for health care.

We thank the outstanding contributors to Academy forums and the sponsors of projects that gave rise to the essays in this volume. The New York Academy of Sciences will continue to provide a forum in which health care policy in all of its dimensions and at every organizational level can be examined, differing views can be aired, and improved policy approaches can be crafted.

The Academy's tradition of independence and commitment to excellence remain at the service of global health care. ■

RODNEY W. NICHOLS
President and CEO

SUSAN RAYMOND, PH.D.
*Director, Strategic Planning
and Special Projects*

1

Institutional Choices

Speaking to the Consultative Assembly of the Council of Europe in 1967, British Prime Minister Harold Wilson observed, "He who rejects change is the architect of decay. The only human institution which rejects progress is the cemetery." Notice, if you will, that Wilson remarked only on necessity; he did not promise ease. Change is always difficult; change in institutions, especially in institutions imbued with culture and tradition, is particularly difficult.

Health care, whether viewed internationally or viewed within the confines of national boundaries, is changing. Disease problems confronted, skills required, knowledge available, technologies on the front lines, forms of payment for services—all have shifted markedly over the last two decades. The institutions that sit astride this ebb and flow of change, however, often find accommodation difficult. Such organizations—whether international agencies, national public health institutions, or the physician's office across the street—often resist the imperative to shift priorities, approaches, and processes to respond effectively to the health care environment. The two essays that follow illustrate how and why health care institutions, one global the other within the United States, might shift their approaches to continue to provide effective leadership into the next century.

Advisory Group on International Health Systems Assessment

The following individuals provided advice and guidance on critical issues facing future cooperation in global public health.

DR. JORGE ALLENDE
Professor of Biochemistry
Universidad de Chile
Santiago, Chile

MS. JOANNE S. BESSLER
Vice President and Chief Human Resources Officer
Booz Allen & Hamilton
New York, New York.

DR. BARRY BLOOM
Professor, Department of Cell Biology
Albert Einstein College of Medicine

MR. PAUL DIETRICH
President
Institute for International Health and Development
Alexandria, Virginia

DR. JULIO FRENK
Executive Vice President
Fundacion Mexicana para la Salud
Mexico City, Mexico

DR. MAURICE R. HILLEMAN
Director
Merck Institute for Therapeutic Research
West Point, Pennsylvania

DR. JOSHUA LEDERBERG
University Professor
The Rockefeller University
New York, New York

DR. STEPHEN S. MORSE
Assistant Professor
School of Public Health
Columbia University
New York, New York

MR. RODNEY W. NICHOLS
President and CEO
New York Academy of Sciences
New York, New York

DR. GEOFFREY OLDHAM
Professor, Science Policy Research Unit
University of Sussex
Brighton, Sussex (U.K.)

DR. SUSAN RAYMOND
Director, Policy Programs
New York Academy of Sciences
New York, New York

DR. GORAN TOMSON
Associate Professor
Dept. of Public Health Sciences
Karolinska Institute
Stockholm, Sweden

Global Public Health Collaboration: Organizing for A Time of Renewal

Introduction

Whether viewed as problems or as opportunities, public health issues will surely represent global priorities well into the next century.

On the one hand, the last decade has witnessed the reemergence of many infectious diseases once thought to be under control, the emergence of new diseases, and the rapid spread of resistant strains of existing disease organisms. No one nation, no one effort, can turn the tide now breaking on national health care shores. Because of expanded international travel, trade, and communications, in addition to denser populations, changes in disease patterns have become, and will increasingly be, every nation's concern. Moreover,

> International public health issues represent opportunities for cooperation that will benefit every nation as well as the global commons. There is much to gain from such cooperation and virtually nothing to lose but illness and death.

despite recent trends toward globalization (spurred by innovations in such areas as telecommunications), forces of disintegration draw strength from fears of lost economic advantage or political power. This pulling apart complicates the global cooperation needed to address shared public health concerns

On the other hand, the advance of science and knowledge now provide, and will provide ever more certainly in the future, the tools to respond to such problems. The opportunity is clearly here to reap significant health care gains through collaboration on research and on disease control. Global public health represents an opportunity for cooperation that will benefit every nation as well as the global commons. There is much to gain from such cooperation and virtually nothing to lose but illness and death.

Yet beyond such broad recognition, complex questions remain: What type of international collaboration will most effectively promote global health into the next millennium? How shall nations and institutions organize to undertake such collaboration? What pragmatic actions will foster greater and more effective global action? This report will offer insights on these critical questions.

To begin, however, clarification of three perspectives is important. What is meant by the choice of "global" as the level of analysis? What is the core focus of discussion of the potentially multidimensional problems of "health"? And what is the time frame of the analysis?

First, what is it about health that is "global"? For the last four decades, "international health" has generally meant "health concerns shared by nations." The designation came to refer not to geographic boundaries but rather to issues of interest or concern to health care authorities or groups from more than one nation. In turn, virtually everything could be called "international," no matter how deeply rooted in the internal problems, decisions, culture, or resources of a particular geopolitical entity. In a variation on this general definition of "international," the term, when linked to "health," came to refer only to issues of poor countries or the poor in industrialized countries. Rather than designating globally shared concerns, therefore, "international" represented the equivalent of "them" in the eyes of the industrialized or wealthy "us."

Second, the "health" in international health has lost its focus over time. Indeed, the charter of the World Health Organization defines health as "a state of complete physical, mental, and social well-being, not merely the absence of disease and infirmity." As a result, "health" can become an ocean teeming with issues and problems of nearly every form and from every societal and economic origin. Again, the programmatic effect at the level of international (and often national) institutions was to

> "Health" can become an ocean teeming with issues and problems of nearly every form and from every societal and economic origin.

expand activities and interests to accommodate the most comprehensive approach to health. Under that umbrella, functions extended from research to services and from training to regulation, within every conceivable sector and discipline. No focus, no meaningful priorities, could be set.

Clearly, then, if the objective is to address pragmatic priorities, actions, and systems for global public health collaboration, some limits must be drawn. It is certainly true that poverty, unemployment, political chaos, war, famine, discrimination, and a myriad of

> Although sharing experiences across nations is often fundamental to problem-solving, many health care problems themselves are ultimately addressed not by the global commons but by national decision-makers.

other factors affect health within nations and across borders. However, it would be difficult indeed to craft actionable ideas for effective global effort from such a complex. The human condition is an unwieldy level of analysis.

Fortunately, especially as global disease patterns change (see below), and as many countries experience increasing economic transformation and rising internal economic capacity, many health care issues are both rooted in, and will likely be resolved (or remain unresolved) through, national action. Nations differ in health status, organization, staffing, and financing because of decisions taken at the level of the nation or decisions taken by even more decentralized political and economic units. Although sharing information and best-practices across nations is often fundamental to problem-solving, many health care problems themselves are ultimately addressed not by the global commons but by national decision-makers.

The focus of this report, then, will be on global public health. The term "global" recognizes that, increasingly, it is not simply nation states that are relevant to health issues, but also the private and non-profit sectors. The private commercial sector plays a critical, cross-border role in innovation that affects and enables national health care. (See Box 1.) The non-profit sector has grown and spread until it now represents a network of services and expertise that transmits health data and service experiences beyond national borders.

The focus on "public health" encompasses both the disease problems and the institutional responses that are "population

Box 1. Global Trade and Global Health

For the past two decades, the increase in world trade has out-paced the rate of global economic growth. Trade is an ever more important component of global prosperity. The extraordinary expansion in global trade has significant implications for global health, both positive and negative. On the one hand, trade in products and services that support and improve health has risen sharply, facilitating access to health care. The value of global trade in pharmaceutical products increased 50% from 1990 to 1994, and the value of global trade in medical equipment increased 30%.

The spread of private corporate presence through trade and investment also provides opportunities for deepening the resources available for public health programs. The Digital Equipment Corporation has initiated a global HIV prevention program with a focus in Asia. Boeing and Levi Strauss have also initiated global programs. In Thailand, multinational and Thai businesses have formed the Thailand Business Coalition on AIDS offering worker education and factory-based intervention programs.

But trade also has its negative effects. Examples of the link between trade and disease abound. Perhaps the most publicly recognizable is recent headline-grabbing anxiety over the spread of "mad cow disease" through trade in meat and animal feed. But the phenomena is quotidian rather than sensational. Expansion in trade within Africa has led to larger internal markets for cassava, resultant shortcuts in soaking and preparation of cassava roots before shipping, and outbreaks of Knozo, an upper motor neuron disease resulting from exposure to cyanide from poorly prepared cassava. Regular reports of infections transmitted by poorly prepared fruits and vegetables are common. But even non-food trade can spread disease. Tourism and migrant worker flows result in the spread of infections such as dengue and hanta virus. Water trapped in products can harbor disease vectors. Outbreaks of communicable diseases in one nation or region can quickly become a regional crisis. The spread of dengue throughout the Americas has become a regional priority with an estimated disease control price tag in Central America alone of nearly $150 million.

How big is the problem? Numbers are elusive in the absence of a sophisticated global surveillance system. However, a recently inaugurated data reporting system, ProMED (see Box 8) logged 300 first reports of infectious disease outbreaks with possible regional or global implications in 1995–96 alone.

> Problems take time to be recognized; institutions change slowly. Hence, it is important to take a long-term view if course corrections are to be taken with sufficient time to navigate the scientific endeavor around vexing problems and toward opportunities now only dimly seen on the global health horizon.

based" and that cross national or organizational boundaries. The focus is not at the level of the individual, but on the problems and interventions that threaten, and therefore require the response of, the global commons.

These two limitations do not provide a perfect solution to the definitional problem. Certainly, sharing of experience and data among nations on, for example, systems for certifying health providers or assessing service quality can be a useful endeavor. Such sharing can certainly be "global" and the subject is certainly "health." Yet because the core of this paper addresses collaboration—i.e., joint action to address shared problems—it will attempt to focus on issues with supra-national implications.

Third, it is important to clarify one other perspective to introduce this analysis. Change is afoot across the globe. Past and current demographic, economic, and disease patterns will create new challenges in the future. The time perspective of this effort is decidedly and purposefully long-term. The problems and collaborative responses to be addressed are those that are anticipated to be important 15 to 25 years into the future. Problems take time to be recognized; institutions change slowly. Hence, it is important to take a long-term view if course corrections are to be taken with sufficient time to navigate the scientific endeavor around vexing problems and toward opportunities now only dimly seen on the global health horizon.

What is the Current State of Global Health?

Several recent studies have addressed in great detail the changes in the disease patterns that have accompanied global economic, political, and scientific changes of the last several decades. This section does not reproduce those

> **The current health profile is characterized by dramatic change, persistent disparities, puzzling transitions, and emerging crises.**

excellent efforts (the bibliography on page 74 lists a range of recent analyses). Rather, it highlights their analyses. A brief review of the evidence is important because it sets the scene for the subsequent formulation of pertinent questions.

The current health profile is characterized by dramatic change, persistent disparities, puzzling transitions, and emerging crises.

Dramatic Change

Nearly everywhere, mortality rates have declined and life expectancies have risen. In general, the last forty years have seen significant improvements in health status, led, in part, by scientific, technological, and medical advances and their widespread application and, in part, by expanded infrastructure, rising incomes, and an increasingly literate public. In much of the world, many people are living longer and more productive lives. Average life expectancy at birth in 1995 was 65 years, and more than 77 years in industrialized countries. By 2020, projected life expectancy at birth will aver-

> **Although a gap in life expectancy between rich nations and poor nations remains, it has narrowed from 25 years in 1955 to 13.3 years in 1995.**

age 71 years worldwide, and will approach 88 years for women in industrialized nations. Although a gap in life expectancy between rich nations and poor nations remains, it has narrowed from 25 years in 1955 to 13.3 years in 1995.

The disease pattern is also changing. Non-communicable diseases such as cardiovascular disease, cancers, and accidents, which have long been the core problems of industrialized nations, are becoming important problems in many less developed regions as well. As life expectancies rise and particularly as populations age, low income and middle income nations will see their epidemiological patterns changing. To date, the process has been slow, but it will certainly continue. By 2020, only in Sub-Saharan Africa are communicable diseases projected to be more important to mortality patterns than non-communicable diseases.

> **The lowest income nations, and those with chronic political instability, continue to fall behind the improvements of most others.**

Persistent Problems

Within change and progress rests a core of persistent stasis. Many past scourges continue. Progress on health indicators is far from uniform. The lowest income nations, and those with chronic political instability, continue to fall behind the improvements of most others. Traditional disease problems—communicable, maternal and perinatal conditions, and nutritional deficiencies—account for 40% of the global disease burden, and nearly half (49%) of the burden in developing countries. Ironically, many of the problems can be addressed effectively by existing knowledge and technology.

The most striking of present disparities are among nations within the developing world. The gap in life expectancy at birth between the least developed countries and other developing countries has widened from 7 years in 1955 to 13 years in 1995. In Africa, particularly, both health indicators and disease patterns share much in common with the patterns of decades ago. For example, while the goal of 80% immunization coverage was reached globally in 1995, twenty five developing countries—9 of them in Africa—report coverage below 50%.

Disparity within nations also persists. The poorest live the shortest lives and are most likely to continue to suffer the consequences of diseases that have ebbed in the remainder of the population. While the pattern is often

> **Tragically, a core of important problems seems to be hardening in global public health.**

most dramatic in developing nations, industrialized countries show a similar disparity between the poorest people and all the rest.

Tragically, a core of important problems seems to be hardening in global public health. Infectious diseases continue to afflict the most vulnerable and impede both health progress and economic prosperity. Ironically, although forty years of effort has shown clear solutions to these disease problems, getting the solutions to those with the problems remains an unmet challenge.

A Perplexing Future

Changes and improvements, coexisting with hard-core problems, create perplexity about the future. The "epidemiological transition" from communicable to non-communicable diseases is not a linear process. There is coexistence and overlap among various types of diseases in all countries, including the industrialized world. Indeed, there is even backward movement as some infectious diseases rise in importance. Over the long term, the trend toward a pattern of change dominated by non-communicable diseases does emerge, and is emerging now in middle, and some low, income nations. How far and how fast this trend will extend depends on a range of complex variables.

As adults become a larger portion of the population, the result is not simply in terms of changing patterns of mortality and premature death, but in changing patterns of illness and disability.

> **A critical element of this transition is the effect of new expectations and changing citizen behavior on the health transition.**

Longer life means not only changes in illness patterns for many countries, but also longer periods of disability. Older people in developing countries will spend longer periods of their older years with disabilities than their counterparts in the industrialized world. A 60-year-old Sub-Saharan African can expect to spend about half of his or her remaining years with a disability compared to just one fifth of those living in established market economies.

Yet even as these trends emerge, traditional problems will coexist on the epidemiological landscape of every nation. Disease problems are not replaced, they are augmented. This augmentation factor is a product not simply of demographics and the natural tendencies of pathogens to prey on humans, but also of poor resource allocation decisions. The massive use of antibiotics, for example, is and will be a great contributor to bacterial resistance and persistent infectious disease patterns.

A critical yet poorly examined element of this transition, concerns the effects of new expectations and changing citizen behavior on the health transition. Throughout much of the world, central planning is being dismantled and markets and societies are enjoying greater freedom. Nevertheless, the balance between private and public poses a dilemma for health care. In all dimensions of life—political, economic, and social—individuals are now the prime decision makers. But how individuals will exercise (or will be permitted to exercise) choices in their health care provision is unclear. How both government and private sectors will seek to influence those choices is also unclear. Understanding how choice will affect anticipated health care outcomes is an important element of future epidemiological patterns.

Changing individual behavior accompanying greater globalization also extends to lifestyle changes that have negative health implications. Changing diets, sedentary lifestyles, tobacco, drugs, and alcohol use and abuse—all are spreading globally and will affect future health patterns in larger and larger portions of the world's population.

It is equally difficult to predict the pace with which countries, particularly the poorest developing countries, will see significant levels of changing disease patterns based on such factors. This is because, in part, it is impossible to predict the rates of long-range economic growth. Major periods of economic stagnation could slow change. Furthermore, it is difficult to predict exactly how the transition will distribute itself among groups within nations. If the long-term economic growth patterns are comparatively robust and stable, but the benefits of growth are not evenly distributed, the appearance of epidemiological change may mask an evermore deeply persistent pattern of traditional infectious diseases among the poor. Nevertheless, it is evident that a significant epidemiological transition is underway throughout the world.

Potential Crisis . . .

While the inevitability of infectious diseases is recognized in the scientific community, humanity in general seems regularly to be caught unawares. The emergence of the human immunodeficiency virus, and the worldwide epidemic

> **Changing diets, sedentary lifestyles, tobacco, drugs, and alcohol use and abuse—all are spreading globally and have negative health implications.**

it has caused; the crisis with Ebola; the resurgence of tuberculosis; the spread of resistant strains of pneumonia—all are reminders that microbes continue to confound humanity's best defenses, spurred on by population growth, rapid urbanization, global transportation, increased trade, and a variety of other dimensions of social and economic progress. The instability of the infectious diseases situation globally—with new pathogens and new outbreaks—has not been a part of the calculus of health planning. Continued changes in the interface between humans, animals, and the environment, together with the natural evolution of the disease agents themselves (and the effects of human action on that evolution through, for example, excessive use of antibiotics), will make it evermore important to begin to anticipate the surprises that almost certainly lie in wait for international health.

> **The instability of the infectious diseases situation globally—with new pathogens and new outbreaks—has not been a part of the calculus of health planning.**

As Richard Levins has noted "Finally, the expectation that infectious disease was in decline [can] be asserted no longer. The accumulation of 'exceptions' and the frustration of our efforts forced a new awareness that diseases rise and fall, evolve and spread and retreat and spread again, and that we have to prepare for a more complex tomorrow than naive progressivism and simple extrapolation would have us anticipate."[1] New, resurgent, and resistant diseases require, in the first instance, an ever-higher level of vigilance at the international level to enable health care professionals to convincingly and continually confront the new and unknown.

> **Surprise can cut both ways. The future will also benefit immensely from unanticipated innovations emanating from medical science.**

. . .*And Potential Opportunity*

But surprise can cut both ways. The future will also benefit immensely from unanticipated innovations emanating from medical science, albeit accompanied by complex questions concerning ethics and resource allocation. Advances in genetics research, for example, will become a factor in diagnosis and treatment. Such advances will also open new pathways for prevention both through applications in immunology and through vector control. The obvious problem will be to ensure that such opportunities are available globally and that the health system capacity to apply innovations is widely distributed.

But opportunity in global health may also originate in other economic and technological sectors. Current and future innovation in information technology and telecommunications present tremendous potential for monitoring and controlling disease outbreaks, as well as for facilitating long-distance training and methods of diagnosis and treatment from remote sites. Increases in levels of education, and the increasingly recognized priority of focusing on literacy among girls, will bear especially important fruit in maternal and child health. The spread of popular communications via satellite transmission provides opportunities for informing and educating the public on health issues and, possibly, affecting behavior. Advances in agriculture based on biotechnology will expand food production. The list is long. There are many good reasons to view the future with cautious optimism.

What Are the Priorities?

W. E. Binkley's biography of Franklin Delano Roosevelt, *The Man in the White House*, contains an anonymous observation about the overwhelming job FDR faced in allocating his time and attention

> **Problems are widespread, resources are limited. Choices must be made. But on what basis?**

to the crises he constantly faced: "It is a mystery to me how each morning he selects the few things he can do from the thousands he should do."

Painful as the process often is, priority-setting is everywhere necessary. Problems are widespread, resources are limited. Choices must be made. But on what basis? No sector is immune from the dilemma; many methodologies have been proffered to facilitate the process. None makes it easy. (See Box 2.)

As noted earlier, the historical tendency has been to address the full range of factors affecting health status and health care in international health programs. Stepping back from that perspective and attempting to be more strategic about priority selection requires a basis for decision-making.

One approach would be to use "value" as the metric, i.e., return on investment. If resources spent at the international level for public health collaboration are thought of as "public international dollars"—as compared to any nation's internal expenditures for individual health care or for public health measures, even if with regard to issues addressed globally, or those similar and very large expenditures in the private sector, (Box 3)—then the question becomes where does the "international dollar" have greatest value? Where is the power of the international dollar greatest in making a global difference, in having impact relative to projected problems? What is the investment strategy that will achieve the greatest return for those resources?

There may be several responses to this query. One perspective would argue that value is greatest where results are most probable and yet least likely to attract domestic financing. This is the global commons approach—the international dollar should be spent on those priorities and in those places where (a) the benefits to all are greatest, (b) individual national action is unlikely because individ-

Box 2. Priority-Setting Methodologies

Few organizations are immune from the discipline of priority-setting. Private corporations, public agencies, and individual wage earners all face the same problem—how to allocate resources to achieve the most relative to their goals. Moreover, most priority-setting requires that judgement about relative outcomes of resources expenditures be made in the face of at least some (and often great) uncertainty. Methodologies for priority-setting also abound.

Environmental policies, for example, have come to rely on the evaluation of comparative risk, scientific evidence about the relative risk to the environment and human health, as a first step in priority-setting. To that assessment is added the cost of reducing each risk and the administrative and political feasibility of taking action. Increasingly, risk assessments are expanded to include the public's perception of risks (which may differ from that of the experts) in a growing recognition that the foundation of effective risk management is public trust and credibility.

A wide variety of other priority-setting methodologies have been tried. Program budgeting and marginal analysis, for example, attempt to place a rational foundation under budgeting choices. More recently in health, in an effort to factor quality of life into priority-setting, Disability Adjusted Life Years (DALYs) have been developed at the Harvard School of Public Health to express years of life lost due to premature death and years of life lived with disability of specified duration and severity.

All priority-setting efforts, however, are based on ethical assumptions about relative values and have ethical consequences. All also, therefore, have political consequences. Priority-setting, by definition, will find some activity, some group, some interest less deserving than another (however that is measured and whatever the ethical base). In turn, controversy and dispute will result. However much managers may try to make priority-setting a technical process, it will quickly devolve into a societal and political debate, often with deep implications for the core values of a people or an organization. Indeed, one of the inherent advantages of open and searching priority-setting is the degree to which it does force an examination of organizational assumptions and values.

ual benefits are not sufficient, and (c) results are most likely.

Yet it must be recognized that the financial imperatives of efficiency are not the only drivers of public health decision-making. Public expectations are also critical: What can science and technology achieve within time frames and policy actions that satisfy voters and taxpayers? Choices must be made on the basis of what actions are effective—what works?—as well as what actions are acceptable—what will the public tolerate?

The delicate balancing act is made more complex by the flow of time. Today's priority will almost certainly be supplanted by a new priority at some point in the future—either because the problems changed, or because innovations made action feasible, or because public acceptance of (or outrage over) problems or solutions fundamentally changed the nature of the priority-setting process. Given such realities, there is no permanent answer in many areas of concern.

An additional element plaguing debate on investments and priorities in health care is the grave problem of ethics. Decisions about priorities are denominated in human life. Money measures do not fully satisfy the concerns of society. At all levels of decision-making—from the bedside to the board room—questions of equity, rationing, standards, access, and quality weave labyrinthine webs of ethical dilemmas around decisions about resources.

There are no prefabricated solutions for priority-setting imperatives. The real need is for rigorous thinking about the questions and then public debate on reasonable lines of answers.

What Problems are Priorities?

Having acknowledged the complexity and fluidity inherent in priority-setting, where should the first and the last international dollar go? In the past, failure to distinguish among problems has led to a fragmentation of international health resources and, in

Box 3. The Imperative of Global Public–Private Partnerships

While the traditions of the last 40 years have led many analysts to consider "global health resources" in terms of public financing, public–private partnerships are critical to any successful strategy. Global public health must build on a multiplicity of strengths across organizations. Partnerships are not one-dimensional. They mean shared benefits, shared costs, and shared accountability. Corporate investments represent nearly half of global pharmaceutical R&D. The overwhelming capacity for turning basic research into viable products, and for mass producing, packaging, and shipping that product, resides in private institutions.

Private sector capacity is also borderless. In all industrial sectors, U.S. corporations now finance $10 billion of R&D abroad, and foreign corporations spend another $14 billion on R&D within the United States. A quarter of the patents issued by the European Patent Office are to U.S. applicants, and another 22% are to applicants from Japan.

The private role is ascendent not just in commercial capacity, but in the non-profit sector as well. The past decade has seen the proliferation of private, non-profit organizations dedicated to health research and service provision. Telecommunications has allowed these institutions to become increasingly networked, with extensive capacity for sharing strategies and experiences. The non-profit capacity is a growing resource for global public health.

turn, the marginalization of many efforts. A plethora of priorities and a limited budget have often meant pennies per capita for each of a constantly rising number of initiatives. Critical mass was rarely achieved in any one initiative.

Beginning at the Beginning: Criteria

What criteria will lead to rigorous choices for the future? In set-

> **At all levels of decision-making—from the bedside to the board room—questions of equity, rationing, standards, access, and quality weave labyrinthine webs of ethical dilemmas around all decisions about resources.**

ting out criteria, it is useful to think in terms of interventions. Where interventions are "good" (i.e., effective, affordable, and socially acceptable), there is the possibility of action. Where they are not good, there is a need for research, be it in the laboratory, in the field, or on the desks of policy analysts. Yet to set priorities for either action or research, problems must be ranked. The difficulty is defining measures of "importance" that can be used to compare problems. How to choose among, or combine, measures such as case fatality rates, differential effects by age and cultural impact, for example, raises not only technical issues but, as noted earlier, often significant ethical dilemmas.[2]

Again, however, priority-setting is made more complex by change. A "good" intervention for an "important" problem today may become a matter for research tomorrow as the problem itself changes (e.g., antimicrobial treatment and consequent drug resistance). Moreover, a problem may be so important in terms of its societal implications that it rises to a level of priority even if neither the interventions nor the available research techniques are "good." Societal pressure can override both scientific capacity and rational resource allocation.

Another critical element among criteria is "leverage." Limited resources can be maximized if they are spent such that they then

> **A plethora of priorities and a limited budget have often meant pennies per capita for each of a constantly rising number of initiatives. Critical mass was rarely achieved in any one initiative.**

attract additional resources to solving the problem or amplifying the scale of the best solution at hand. The likely resource multiplier effect is an important consideration in choosing among alternatives, especially when public resources are scarce relative to the total magnitude of a problem or opportunity. Moreover, spending at the international level can leverage national investments. Creating international programs and leadership provides incentive for national attention to data collection or prevention programs as well as motivation for national leaders to act on problems or opportunities because recognition will be international.

> **Creating international programs and leadership provides incentive for national attention to data collection or prevention programs as well as motivation for national leaders to act on problems or opportunities because recognition will be international.**

Problems

Global public health is faced with many complex problems. No single organization can address all these problems, perhaps not even most of them. But there are some problems for which solutions are available or are on the horizon. Some of these are serious enough to be worthy of immediate, aggressive attention. Looking 25 years into the future, several key trends suggest the most compelling priorities for global collaboration.

The first is infectious diseases control. Unlike many health problems that can be addressed at the national level, infectious diseases cross borders. Their control requires fundamental collaboration among nations and regions. This is even more important as new, emerging, and resurging infectious diseases gain a greater foothold in the human population, and as the growth of dense urban concentrations in low and middle income countries pro-

> **The ability to track and assess raw information through large data bases, to analyze trends, and to create global health intelligence out of a constant flow of information is critical to the success of global health collaboration strategies.**

motes disease transmission. It will also be increasingly critical to anticipate surprises, to develop scenarios based on differing assumptions about the risk of surprises, and to plan for an effective response.

A second clear priority is the capacity to prevent and/or respond to public health problems. This capacity is critical at two levels. Within nations, public health systems need sufficient depth and stability of human resources and infrastructure to consistently identify and address priority problems. Among nations and private institutions, a capacity to rapidly communicate and coordinate is a prerequisite for effective concerted action. The latter capacity is of particular importance from a global perspective. The ability to track and assess raw information through large data bases, to analyze trends, and to create global health intelligence out of a constant flow of information is critical to the success of global health collaboration strategies. The capacity problem involves both human resources and information systems, and can only be addressed collaboratively through mutual consultations, pooled resources, and international institutions.

A third priority stems from the implications of changing demographics. As fertility declines and population growth slows, populations are aging around the world. More people are reaching older ages. Increasingly, the fastest shifts in age-

> **Global health organizations can (and have) become balkanized by internal budgeting competition that pits, for example, malaria versus tuberculosis versus other diseases, one by one.**

structure are being experienced in the middle-income and low-income countries where the proportion of population aged 65 and older is projected to increase by 200% or more between 1990 and 2025. This trend will also shift the pattern of disease burdens, the level and type of public health infrastructure that will be needed to prevent and respond to those patterns, and the nature and level of financing that will be required for health services. Although non-communicable diseases do not pose cross-border threats in terms of direct transmission, global influences are important. Non-communicable diseases may not cross borders, but their risk factors can and do.[3] Trade and communications can and do change tastes and lifestyles which impact health, especially for diseases related to behavior such as tobacco or alcohol use. Consequently, all nations will benefit from an increased understanding of non-communicable disease mechanisms and public health prevention strategies.

What Functions are Priorities?

Having addressed disease or health condition priorities, one then asks what functions have most value at the international level. In fact, resources are often not spent on problems, they are spent on the programmatic functions that drive toward a solution. For example, research, training, and public education are all "functions" that provide alternative (or complementary) approaches to a specific disease control problem.

Thinking in terms of functions unifies substantive priorities. Where priorities are many and appear to be competing, functions can identify cross-cutting, common strategies. Global health organizations can (and have) become balkanized by internal budgeting competition that pits, for example, malaria versus tuberculosis versus other diseases, one by one. One may not need to trade off among infectious diseases in seeking priorities, however, if general surveillance is targeted as a functional priority. While a functions approach does not obviate all difficulties in disease or health condition priority-setting, it may introduce a way to counteract fragmentation and seek common themes, which in turn provides a centripetal force to strategic planning.

But functions are as plentiful as problems. One need only look at the existing range of programs in international health to appreciate the dilemma. "International health" within a single institution can encompass drug procurement, service provision, regulatory advice, research, advocacy, training, standard setting—the list is long. Spreading limited resources across multiple functions incurs the same price as problem proliferation—resources are dissipated across a wide spectrum without allocating sufficient resources to any one effort to make a difference. Again, choices are necessary.

Before suggesting several priority functions, however, it is important to be clear on context. The discussion that follows focuses on the normal course of global public health collaboration. But it is also clear that things are not always normal. Emergencies abound. Famines, wars, natural disasters—all create spikes of crisis on the public health radar. It is clear that an institutional framework is needed to monitor and respond to the public health implications of such crises. Crisis management in the form of emergency humanitarian programs is an important function, but it lies outside the focus of this report.

There is a core set of functions that, it is widely agreed, provide a key foundation to collaboration in global public health. These functions represent the central shared concerns that are in the interests of all nations to secure, but whose pursuit no single nation can ensure.

> No one nation, no one region can undertake effective action in isolation. Surveillance requires assessments within and across borders, and the sharing of data among nations and regions.

Individual nations may (and must) carry out these functions, but whatever is done individually, success requires that some elements of effort must be pursued jointly.

The first such function is monitoring and surveillance. Rigorous assessment and verification of changing disease patterns is essential for effective prevention and control and for resource allocation decision-making. Within nations, of course, public health monitoring of disease is a critical basic function. Whether concerned with drug abuse in New York or AIDS transmission along trucking routes in India, national health requires deep capacity in monitoring and surveillance. The importance of that function is not always recognized, even in the most industrialized nations. In the United States, for example, 12 of the 50 states have no professional position responsible for the surveillance of food-borne and water-borne diseases.

Yet increasingly, no single nation and no single region can undertake effective action in isolation. Surveillance requires assessments within and across borders, and the sharing of data among nations and regions. Events in one nation can quickly have consequences in others. As a function, surveillance requires collaboration on systems and standards for information that will transform data into "intelligence" that can be acted on; infrastructure for data collection and analysis at the regional and international level; and development of the cadre of well-trained professionals for both detection of and response to unusual disease occurrences.

Surveillance also applies to health systems. The ability to respond to health problems is grounded in the capacity of national health systems. Surveillance of the status of and changes in technology, financing, and system capacity at the level of national and regional health systems can provide important guidance for developing effective strategies for strengthening international collabora-

> **Both the necessity for research, and the capacity to carry out research, are increasingly international. It is in the interests of the global commons to strengthen the capacity of the research community in the developing world.**

tion in disease control.

Moreover, health system monitoring must also be interdisciplinary and intersectoral. Health systems do not live on a mountain top. They are affected by the pace and cycle of change in education, employment, housing, and transportation. Developing methods for system monitoring and surveillance that factor in interventions from non-health sectors is a difficult problem for which there is no ready analytic solution.

The second generally accepted priority function is information dissemination, both in terms of best practices, innovations, and technology, and in terms of training. Information about how to improve health care provision, develop preventive programs, manage resources, and the like is born of national experience. Without organized and continuous professional effort, such experience and information will not necessarily flow to other geographic areas. Moreover, it will not flow automatically from the scientific community to the political/policy triggers that must understand, accept, and act on the information. Finally, and perhaps most importantly, it will not flow to those providers and public health practitioners "in the field" who are closest to and responsible for health interventions. Proactive efforts are needed to guide the flow and use of information.

Yet it is in the interests of the international community for the capacities of every nation to be adequately developed to respond to priority health care needs, especially those that affect the global commons. Hence, collaboration in information dissemination is a growing priority that requires institutional leadership at the international level.

A third priority function is research. As noted earlier, where existing interventions are inadequate for either technical, cost, or societal reasons, priority health care

> **The "research" function is not only a matter of understanding pathogens, it is also a matter of analyzing policy systems.**

problems cannot be effectively addressed. Research is essential. Many nations do and must support deep life sciences and disease control research. Where national interests converge and global public health is at stake, global support is also critical. Indeed, the research process has itself become more international.[4] The globalization of research has its roots in the rise of high-capacity, rapid modes of electronic communications; the expanding scientific capacity throughout the world; and the rise of new scientific problems that are global in nature and widely acknowledged as common priorities among scientists and nations. Both the necessity for research, and the capacity to carry out research, are increasingly international. It is in the interests of the global commons to strengthen the capacity of the research community in the developing world, to collaborate with global efforts in shared disease priorities, and to ensure that major diseases endemic to developing countries can be analyzed and addressed in their epidemiological context.

It is important to recognize that the "research" function is not only a matter of understanding pathogens, it is also a matter of analyzing policy systems. The research agenda includes the development of an understanding of the operations of health systems and the "political mapping" that makes policy possible. Both systems and policy understanding is a crucial element in designing public health interventions that are effective on the ground.

Beyond this core of functions, the terrain is hazardous. There is no clear consensus on what functions an international health system should undertake. Indeed, there is deep disagreement. In large part, this may be a result of differing views about international organizations. Some argue that international organizations should undertake only functions that nations cannot themselves carry out. On the other, some argue that the central function of international organizations is to redistribute resources among nations. Clearly, the functions of international health collaboration would be very different for each of these two groups, as well as for the wide range of combinations of views that rest along the spectrum between them. Nevertheless, despite differences of economic and health status, all nations will benefit from a global approach to the critical public health problems that impede progress and endanger health within nations.

Current Institutions and Resources

A wide variety of public institutions are engaged in various dimensions of global public health collaboration. In some cases, global health issues are their core mission, (for example, the World Health Organization). In others, health is a portion of their global or regional project portfolio, (for example, the World Bank and the regional development banks). In still others, such as the national agencies charged with disease surveillance, global public health concerns are a pathway to core disease prevention and control mission within a nation.

Private institutions are also critical actors. The private commercial sector possesses the world's capacity to apply scientific progress to innovations that will reach those in need. The private sector represents half of the research and development capacity in the global health enterprise. Private foundations have also been important historically in the evolution of global public health strategies. While many foundations have scaled down such involvement, many others remain committed to global health issues. The private non-profit sector also plays a critical role. Non-governmental organizations span national boundaries and bring capacities in both research and program implementation to and among nations. Private non-profits often provide the networks through which nations exchange information and experience.

Appendix 1 provides a summary of the leading current institutions on the global public health stage, and an estimate of their resource commitments, totalling around $10 billion. Box 4 provides an overview of key data.

Global Public Health Collaboration in the Future

Over the past five decades significant strides have been made in disease control. Many "vital signs" of the global corpus are moving in positive directions. Yet old challenges remain, and new challenges certainly lie ahead. Planning for the future, therefore, must not simply be an extrapolation of the past.

Given the substantive and functional priorities, and based on existing experience with organizing for resource allocation to international health, there is room for new, creative thinking. Often, such thinking is aided by "zero-basing" the analysis. Taking a totally fresh view of what is possible and what may be optimal, unencumbered by the constraints of existing programs, commitments, or organizational stakes, can help to drive toward new insights.

A Global Public Health Resource Exercise: What If You Had a Billion Dollars?

To spark creative thinking, assume that there is made available $1 billion per year in constant dollars in perpetuity for global public health cooperation, to be allocated to cross-national efforts at

Box 4: International Health Funding

Drawing a full and complete picture of public and private expenditures on currently defined "international health" is difficult given the wide range of bilateral, multilateral, private, non-profit, and foundation programs on the scene. Annual expenditures probably total at least $10 billion for programmatic action. While significant, these resources are only a minor portion (perhaps 10%) of total annual resources allocated to "development assistance" globally. Moreover, the "international" total pales in comparison to total expenditures of perhaps $3 trillion on health care within nations. Summaries of the expenditures and program categories of the major institutional actors are contained in Appendix 1. These resources include:

World Health Organization: WHO's total budget is about $1 billion per year, 46% of which represents the regular budget of the institution and 54% of which represents "special programs" that are targeted at particular priorities (e.g., tropical diseases) and separately funded by interested nations or consortia of donors such as the World Bank, the Rockefeller Foundation, and bilateral aid agencies.

World Bank: "Population, health and nutrition" lending commitments average about $1.2 billion per year, and have risen from 2.5% of Bank lending in 1989 to 10.9% in 1996. Total PHN disbursements have risen from $143 million in 1989 to nearly $1 billion in 1996.

Regional Banks: The Inter-American Development Bank lends about $1.5 billion for health and education annually. The Asian Development Bank does under $50 million per year in health lending, and the African Development Bank in 1994 made loan commitments of about $34 million, or 2% of its lending.

UNICEF: UNICEF's annual budget is just over $1 billion, with 36% allocated for procurement of medicines and vaccines, 52% for grants and technical assistance for health, education, and nutrition, and 12% for administration and other program support.

Bilateral Aid: Bilateral development assistance programs of individual nations also have significant health components. The Untied States' program, for example, allocates about $1 billion per year for population, health, and nutrition assistance. Together, nations of the European Community provide an estimated $5 billion in health assistance through their bilateral and cooperative aid programs.

Other Bilateral Civilian Programs: Individual nations also allocate domestic financing to programs with direct international health implications. The National Institutes of Health, for example, allocates $186 million explicitly to international programs, but its research grant allocations for several tropical diseases are multiples of those of WHO itself.

Corporate Commitments: Private corporations, particularly the pharmaceutical industry, are key assets in the global health system. Pharmaceutical companies account for 44% (about $24.7 billion) of annual spending on health research worldwide.

Non-Profits and Foundations: Particularly important in the development of strategies and networks and the provision of goods and services, non-profits and foundations span national boundaries and are increasingly the agent through which information and innovation flow.

---

the level of the global commons. What would be the most important, most productive investment for those resources relative to health priorities? What would be the most productive use of those resources relative to functions? Keeping in mind the marked changes in national capabilities over the past generation, what organizational options or principles would best ensure that this international resource allocation would be carried out with the greatest efficiency?

> **Planning for the future must not simply be an extrapolation of the past. Many "vital signs" of the global corpus are moving in positive directions. Yet old challenges remain, and new challenges certainly lie ahead.**

Surveillance

With regard to disease surveillance, there is a need to rebuild or restore, and then maintain, technical infrastructure. This requires attention to field epidemiology, regional and reference laboratories, the clinical base for laboratory capabilities, and a "swat team" capacity at the international level for rapid response to findings that appear to signal outbreaks or the emergence of new agents.

For health system surveillance, resource allocation would support data reporting and analysis systems that would enable monitoring of flows of health technology and resources among nations and regions. Both disease and health system aspects of monitoring and surveillance require concerted attention to capacity-building in many regions of the developing world.

Research

Focussing on research, the necessary capacity within a global health institution is the ability to identify the needed research on key priority problems. The research function envisioned would not actually carry out research on, for example, fertility technologies or new vaccines within the international organization itself. It would not finance its own research laboratories and personnel. Rather, it would seek to leverage the investments in such capacity within other private and public institu-

> **The proposed research function would not actually conduct research, but would seek to leverage the investments in research capacity within other private and public institutions.**

tions. It would have the internal capacity or external networks necessary to make decisions about priorities and productive research. It would, in effect, look for and fund the "best buys" in research relative to public health priorities, the state of scientific knowledge, and the research investments of other, particularly private sector, actors.5

Information for Use

For surveillance and research, moreover, accumulating data and using information are two very distinct processes. One of the central problems in matching global health functions to action is that the locus of responsibility for acting on information is often not at the global level. Instead, actions must

> There is a need for skepticism about traditional views of public health choices and priorities and a need to advance the understanding of how to set priorities for international collaboration.

be organized and implemented at the national or even sub-national level in both the public and private sectors. The "national versus international" debate over public health problems of cross-national concern is often a false dichotomy. The two are not alternatives, but complements. Global cooperation to generate information and national action to use information should create a public health continuum. However, the process of carrying out the critical functions at the global level, then, must also include strategies to create self-interest and capacity within the national "stakeholders" to that information so that action follows insight. There is no simple solution to this problem; national and local policy-making (and politics) are often resistant to global advocacy. Nevertheless, global health leaders do bear the burden of being global opinion leaders in encouraging national action backed by research and compelling evidence on the benefits that flow therefrom.

Better Ways to Set Priorities

Additionally, there is a need to allocate resources to improving the process by which priorities are set. The many tensions in international health, the various sets of values at work, the range of nation-states involved (the stakeholders), the complexity introduced by rapid change and unanticipated surprise, all argue for greater attention to the "how" of decision-making, not just what decisions to make. There is a need for skepticism about traditional views of public health choices and priorities and a need to advance the understanding of how to set priorities for international collaboration. For example, should priority be placed on acquiring new knowledge to anticipate surprise, or on applying existing knowledge to overcome recognized problems? "Scenario building" based on clear and explicit risk assessment must be a part of ongoing priority setting, reassessment, and change. Such rigorous approaches are not widely used in international health decision-making, but they must become sharply honed tools if international health collaboration is to keep pace with global change.

Getting the Real Budget

Budgeting a hypothetical billion dollars is, obviously, fraught with two key unknowns. First, what are reasonable unit costs? How much would one more surveillance unit cost? Is there an accurate price tag for the next malaria research breakthrough? Second, even if the unit costs were known and the necessary level of effort could be estimated, is there absorptive capacity in the global (or national) public health infrastructure for those investments? Could the money be spent productively? How long would it take to build capacity to do what is desirable?

Consider, for example, if one felt that the surveillance function was so critical and so international that it should receive a third of the billion dollars. Is that realistic? Can $300 million or more be spent productively every year for the next ten years (i.e., $3 billion) on surveillance? Would what is needed to improve international collaboration in surveillance require that much money, given the current existence of a foundation upon which improvements can be laid? Box 5 provides an example of what significant improvements in surveillance might cost, and what the payoff might be.

Similarly, if research is the critical function, how much money would make a difference? The extra resources to be spent are not the first dollar allocated to research, they are the next dollar and that might affect the anticipated productivity of the investment. Is there half a billion dollars of collaborative, productive research that is not now being financed? Can $5 billion worth of productive effort be accomplished over five years?

And in what time frame? This is a critical issue. A recommitment to research may require a 25-year projection of the need for more capacity; people take time to train; laboratories take time to build. Unless the target is clear, near-term budget commitments can easily be off the long-term mark.

Or put another way, how much would effective information dissemination and training really cost? Electrons are relatively inexpensive. Would $100 million per year over ten years create a sustainable capacity for an efficient switching mechanism to ensure that new knowledge, insights, and innovations are broadly distributed and, when distributed, can be applied in a range of national conditions? Is it too much or too little, given the recurrent costs for staffing and skill/systems upgrading as information systems continue to evolve?

There are no simple answers, but zero-basing the discipline of the "billion dollar budget" imperative does make the question clearer. Box 6 provides an example of how such a sum might be allocated given the priorities set out in this report.

Box 5: What Would It Cost to Upgrade Global Surveillance?

There is widespread recognition that the global system of infectious disease surveillance is under significant stress. At the same time that both telecommunications and computing technologies are opening up deeper and faster capacities for data analysis and transmission, and as human resources capacities in the developing world are growing and hence enabling more complete disease monitoring, investments in surveillance infrastructure lag behind. While there is no detailed estimate of the likely costs of establishing and operating a more complete series of surveillance laboratories, a reasonable estimate can be derived from the current "rule of thumb" based on current costs. A surveillance laboratory that served as an analytic reference site for health systems throughout its catchment area might require staffing by two senior and four junior scientists. Together with start-up costs and operating supplies, the unit might cost about $2 million per year, plus or minus 10% depending on location. A network of 10 laboratories throughout the world would then cost $200 million over the next decade.

Again, using an appropriate rule of thumb, the core repository for network data, as well as a central laboratory capacity to assess samples referred by network labs might cost another $2–4 million per year, adding between $20 and $40 million to the total system cost. So, for an expenditure of about $250–300 million over ten years, the global surveillance capacity for infectious diseases, including new and emerging threats, could be significantly improved.

While this total is considerably more than is now being spent on surveillance (e.g., the total budget of the Special Pathogens Branch of the U.S. Centers for Disease Control is $5.5 million), it would mean less than a one percent increase in the total amount of government money now spent worldwide on health research ($28.1 billion). Cast in defense terms, this $300 million is the disease intelligence budget to gauge the disease "threat" to global public health.

Organizing the System for Global Collaboration

The past several years have seen much discussion of the role of international organizations in the new geopolitical context, and of the necessity of change. Much of that discussion has been accompanied by alterations at the organizational margins—a directorate move here, a staff slot eliminated there.

However, the overall global public health collaboration system is at the end of productive incrementalism. Squeezing another penny out of an institutional budget will not create the innovative change needed to respond to tomorrow's needs and opportunities. What is the key organizational profile against which change should be measured?

Today, virtually every organization must think and organize globally. The end of the Cold War, the spread of open markets and open societies, the relentless pace of competition, and the blossoming of electronic communication has made a multinational of even the smallest of businesses, associations, or non-profits. Hence, much is being learned about how best to organize and manage institutions that must act globally. The rules of organization and management are being rewritten, and transformational thinking permeates the management literature. Within this new thinking, several principles are becoming increasingly accepted as key to organizational efficiency and effectiveness.

Partnership.

Faced with a world of speed and complexity, a global international health organization cannot expect to carry out all functions for all problems at all times. It must seek out and nurture the capabilities of all potential players—public, academic, commercial, and non-profit. A successful global health organization will need to transcend traditional boundaries and sectors, embracing the best, most efficient capacity wherever it is located. This

Box 6. A Billion Dollar Budget

If you were guaranteed a billion real dollars per year for ten years for global public health, how would you spend them? What functions would occupy pride of place? Based on the analysis in this report, distribution of a billion dollars would:

- be dominantly allocated to disease surveillance. Perhaps 25% of the total would be reserved for this purpose, with an emphasis on developing long-term personnel capacity and global monitoring and response systems.

- favor information dissemination and capacity building as a second priority. Perhaps another 20% of the billion dollars would be allocated to this priority, with an emphasis on creating the virtual networks that would promote widespread reporting on diseases and on health systems, as well as improved communication of this information and its importance to decision makers.

- allocate about 10% of the funds to research, with a focused emphasis on those areas not being addressed by national research programs. The research program would be a peer-reviewed, merit-based grants system.

- recognize the instability of politics, geology, and atmospherics, and reserve 10% of funds for crisis response.

- reserve about 15% of the budget for miscellaneous functions such as global public health education but with an earmark for an experimental program on research, modeling, and simulation to improve methodologies for priority-setting.

- firmly cap central administration and overhead to no more than 20% of the budget. In carrying out these priorities, the emphasis would be on expending resources in decentralized personnel units.

process of organizational partnership must itself be capable of adjusting to new threats and new opportunities. In short, the global health organization must be built to change.

> **Faced with a world of speed and complexity, a global international health organization cannot expect to carry out all functions for all problems at all times.**

Decentralization and the End of Geography.

Organizational decentralization is a key element of efficiency for global institutions. Getting the product or process (or data or research) closer to those who need or use it ensures that responsiveness is built into organizational structure.

Moreover, the decentralization is not just of organizational capacity, but of function. The key issue is not just where management authority rests, but where it rests to do what. Decentralized units are not miniatures of headquarters. Their functions represent a critical mass of their responsibility for serving their markets. Headquarters does not play a duplicative role. Rather, headquarters is the "switching mechanism," integrating system-wide planning and providing the point through which communication flows and by which "lessons learned" are spread throughout the system. For global health, such decentralization of function (as well as organization) would place virtually all implementation authority in the field.

Traditionally, "decentralization" has meant organization by geography; organizational sub-divisions followed regional geography. However, as economic development proceeds it washes over regions with differential effects. Brazil looks more like Singapore than it does Peru. Morocco shares more with Portugal than with Libya. Yet most organizations are still managed by geography. Among those changing most quickly are global financial institutions, which find it most efficient to organize their human resources skills and financial offerings by the capacity of their consumers and customers. "Emerging markets" divisions combine countries from Brazil to India, from Taiwan to Hungary. Breaking with geographic organization allows a bank to bring the correct mix of multiple dimensions of financial services to client institutions. Hence, services are neither over- nor under-engineered for the customer. Rather than duplicating technical capacities in national or regional markets, cross-functional teams of specialists and managers focus on organizing resources to respond to the categories of customer needs, wherever they are located.[6]

Ridding global health systems' organizational thinking of its geographic roots would require a focused emphasis to build on and respond to capacity. The "decentralization" would not be geographic, but on the basis of grouping nations according to their capacity for surveillance, information use, and research; tailoring systems to meet and build on that capacity; and then, importantly, regularly monitoring that capacity and re-grouping to reflect evolution and growth. And this formula, too, would change over time as disease patterns change and countries evolve.

The core problem with decentralization is how to create systems that allow central decision-makers to maintain navigational control while delegating authority outward. New management thinking recognizes that, in fact, central managers often do not have the expertise to make key operational decisions, especially in organizations premised on science and technology. Organizations must move decision-making incentives out to decentralized units. Headquarters sets the strategy; decentralized units take full responsibility for implementation.

However, with decentralization comes the ever-critical element of accountability: decision authority is accompanied by responsibility for outcomes. In private organizations, such an approach is possible because it is accompanied by rewards and sanctions. Accountability measures are clear; the market measure of merit is clear. Employment implications are equally clear: jobs are at stake unless performance is high. In global public health institutions such elements would need to be developed.

Most global organizations are still experimenting with the decentralization dilemma. In some cases progress has been made. Both functions and personnel are organized into networks of expertise in the field. These are combined and recombined into problem-solving teams by a lean headquarters staff that serves as navigator for a fleet of activities. Asea Brown Boveri, for example, operate a global $25 billion technology and energy company of 210,000 employees with a headquarters staff of only 100 people. While private corporate organizations cannot provide a perfect model for public systems (where governments can—and often do—abdicate public health responsibility, dedicated management teams may be scarce), "lessons learned" can provide important organizational insights. (See Box 7.)

Information Systems: Toward Virtuality.

The premium in global institutions is on information technology systems. But the critical insight of new management approaches is the recognition that information systems are not simply pathways between central and field management nodes. Indeed, when they are conceived and operated as mechanisms for management command and control they impede organizational success. Rather, they constitute the critical web that links the all-important decentralized units to one another. Information systems are seen as less a means to harness data for "reporting" to management, than a way to diffuse knowledge and provide feedback among all management nodes. In the most effective global organizations, information technology is a means to link information sites to one another so that people with information can share it and create knowledge. This process also affirms the active relationships of all "parts" to a larger effort.

Given the importance of the data and information element of all global health functions, sophisticated and constantly evolving

---

Box 7. ABB: In the Forefront of Global Corporate and R&D Management

Asea Brown Boveri (ABB) is widely regarded as a pioneer in global management among technology-based corporations. Headquartered in Zurich, ABB was formed in 1987 through the merger of the Swedish industrial flagship Asea and its Swiss equivalent Brown Boveri. ABB has pursued a policy of creating a "multidomestic" enterprise, i.e., an organization that is global but with deep roots in every nation. Functions that are common are shared globally to achieve cross-border economies of scale, but deep local roots are nurtured in every country or market with close working relationships with local universities and regional governments. The management emphasis is on the creation of networks. ABB earns $36 billion annually, and employs 215,000 persons around the world, yet operates with only 100 staff at its Zurich headquarters.

ABB extends its managements strategy to its $2.3 billion R&D operations. More than 90% of its R&D operations take place within the networked business units. The remaining 10% at the corporate level is also decentralized to six laboratories in different countries, which individually bring specialized expertise to bear on a focused series of 15 integrated research programs. A key technique for ensuring that the highly decentralized R&D program remains linked to overall corporate goals is job rotation between R&D and operations staff. Every R&D team includes qualified (usually Ph.D.) managers from corporate operations, and individual scientists are posted in operations for between three and twelve months to better understand the problems and needs of particular technology products or national markets.

information systems are essential. With regard to the priority problems of surveillance, information dissemination, and research, there is a need for a "switching mechanism", a strong communications and convening function, including electronic capacity. The international institution must have the capacity to identify and rapidly disseminate the most innovative approaches, the newest findings, and the most informative data.

That capacity has organizational implications. The information technology and telecommunications revolution has taken organizational development to the edges of virtuality. Networks of people create and share large databases, adding value to raw information by joint analysis and the development of collaborative programs for experimental implementation—all without a physical location. Conceptually, virtuality in organization is not new. A corporate sales force often operates without a physical gathering point. Journalism is comprised of correspondents and reporters who do not meet in one physical place. The Open University in Great Britain does have a physical home base, but only for administration. No students and few faculty are to be found there.

In the future, technology is likely to enable a higher degree of virtuality in many organizations, particularly those for which value is to be found in information and information services rather than in physical products. Virtual organizations provide a means to enhance value with efficiency. (See Box 8.)

On the other hand, virtual organizations create their own management problems. Management in many traditional, large organizations is premised on oversight and control. How does a manager oversee those he or she cannot see? One of the key elements is trust. Successful organizations that maximize technology for the purposes of decentralization and efficiency must trust that those tied into the organization share the organization's goals, priorities, and values. Virtual organizations must be organizations of colleagues; they are self-policing through a professional ethic. In turn, individual failure to uphold the integrity of the organization must result in the departure of the individual, or the entire basis for the organization is at risk. There can be no guaranteed sinecures. This makes virtuality difficult to apply in many far-flung organizations pursuing a wide range of diverse goals and programs, especially where national politics are a factor (as they are likely to be in any global health institution that relies on national governments for resources or governance). Nevertheless,

---

Box 8. A Virtual Organization

Many organizations are being created with little in the way of bricks and mortar. Electronic communications and widespread access to technology, along with decreases in the prices of such systems, has encouraged this trend. Traditional structures and bureaucracies are not compatible with the speed of analysis and decision-making possible (and often necessary) in the information age. In the private sector, the "corner office" as a measure of management success has disappeared. Bell Atlantic, for example, expanded the number of its managers operating almost solely through telecommunications from 100 in 1991 to 16,000 in 1994.

In Canada, the Canadian Network for Technology-Based Learning is a nationwide research network of universities, industry, and government researchers whose goal is to facilitate and support research partnerships in technology-based learning that will enhance Canada's competitive and social/economic potential. The Network and its projects, have no "home" of bricks and mortar, but exists and operates electronically. Recently, a similar suggestion has been made for establishing a global network Program to Monitor Emerging Diseases (ProMED). ProMED would act as a electronic linkage among existing strategically located institutions in the developing world, which could act as sentinel centers for disease surveillance. The program would be experimental, but focused on the most important endemic and emerging diseases and the most vulnerable areas. The ProMED network would be focused on recognizing unusual outbreaks at an early stage in order to prevent their spread, relying dominantly on existing (albeit improved) infrastructure to implement necessary actions.

---

virtuality could be a critical element of certain highly technical and data-based organizations, such as those that address health surveillance.

A critical element for virtuality is the development of strategic alliances. Organizations specialize where they bring high value added

> **The information technology and telecommunications revolution has taken organizational development to the edges of virtuality.**

and form alliances with other specialized organizations. Both some of the largest and some of the newest technology-based corporations have little manufacturing capacity at all. Products are designed and then sent to production specialists for manufacture. Indeed, a 1995 Financial Times survey stated that firms with strategic alliances are 50% more profitable than those without.

For a global health organization, the "high value-added service" may be in concept design and data management. Strategic alliances with other national, regional, or specialized organizations allow the global organization to accomplish surveillance, research, or the like, without actually investing in the human, infrastructure, or systems capacity needed to carry out the task. The global organization becomes the head, but the arms and legs can belong to others.

Human Resources.

Where information and technical capacity are the core of value, human resources become the organization's most valuable tangible advantage. Attracting and keeping the best human resources is one of the most critical tasks for a technical organization, because, in a very real way, the organization's key assets walk out the door every night. Moreover, when the greatest value is in human resources, the contract between the organization and those resources must also reflect that value. In high-technology enterprises, software programmers and hardware engineers do not consider themselves

> **Strategic alliances allow the global organization to accomplish surveillance and research without actually investing in the capacity needed to carry out the task. The global organization becomes the head, but the arms and legs can belong to others.**

"employees" or "staff"; they are members of the firm, with an interest and a stake in its strategies and its growth. Similarly, in global health organization, the importance of information and hence the need to attract and keep the highest quality human resources will require that the people involved be deeply committed to and involved in the future of the organization.

Finding and keeping the best people has always been a problem for public international organizations. But in the future it may be even more complex. Forty years ago, most private organizations did not think globally, and hence highly skilled human resources with global interests gravitated to international organizations. Today, all organizations, private and public, spread international wings, and public international organizations have to compete for even the most devoted internationalists.

In global health, the human resources element of organizational design will require great attention to capacity building in the developing world, both for the purposes of creating truly integrated global strategies and for the purposes of strategy implementation. The "head" of global strategies will be only as effective as its "arms and legs" in the field.

The Steering Mechanism

For all global organizations, engineering a responsive steering mechanism is a recurrent challenge. The first dilemma is developing adequate evaluation systems that produce meaningful measures of progress and developing the capacity to use those measures to plot and change course. The more decentralized and virtual the organization, the more complex the problem. How do you know when you are doing well? How do you know when you are done, when investments in one area (e.g., immunizations) could more fruitfully be redeployed in others?

> **Today, all organizations, private and public, spread international wings, and public international organizations have to compete for even the most devoted internationalists.**

First, evaluation systems. There is a need for a manageable set of clear indicators for each core endeavor. Direction must be measured by these indicators, not by the mere presence of last year's budget. The fact that a program had a $100 million budget last year (or any budget at all) should, in and of itself, provide no guidance for the level or priority of its allocation next year. The measure for budget allocation decisions is the level and nature of the problems relative to other problems, and the demonstrated evidence that the resources can and do make a palpable difference. In this way, sophisticated evaluation systems will not only provide evidence of progress, but will ensure that results constantly change the agenda of the organization itself.

Second, management must be committed to constant flexibility and change. Leaders must expect that next year will not look like this year. Where leadership becomes trapped by old wisdoms, new opportunities and changing circumstances are often missed, particularly where waters are relatively uncharted. Leadership must be accountable for progress made toward the horizon not merely the distance traveled. In turn, that requires that the selection criteria

> **Leadership must be accountable for progress made toward the horizon not merely the distance traveled. That requires that the selection criteria for leaders in global public health institutions be weighted heavily toward change management skills.**

for leaders in global public health institutions be weighted heavily toward change management skills.

Accountability raises a final organizational problem—governance. Although it is often seen as a problem only of international public institutions (witness the complexity of debate over reform of the United Nations organization), governance is a rising problem for many private international institutions for whom a sense of nationality is elusive. With operations in 160 countries, components sourced in every geographic region, assembly operations in Peoria and Panang, service centers from bush to barrio, international private corporations also must struggle with new methods for making decisions, developing representative boards of directors, and ensuring adequate information flows. Global public health institutions should reach deeply into the emerging experiences of private organizations for fresh ideas on how best to ensure that governance structures, even within the constraints of nation–state representation, promote organizational flexibility, and insist on world-class leadership. ■

Notes

1. In *Disease in Evolution: Global Changes and Emergence of Infectious Diseases.* Annals of the New York Academy of Sciences, 740, 1994.

2. Murray and Lopez, eds., *The Global Burden of Disease* (Cambridge: Harvard University Press, 1991), incudes a searching discussion of some of these issues.

3. Julio Frenk, *et al.* "The New World Order and International Health." *British Medical Journal* 314 (May 10, 1997).

4. For a general discussion of this point, see *Global Cooperation in Science, Engineering, and Medicine,* a special report from the New York Academy of Sciences' Science Policy Program, February 1996.

5. The report of the Ad Hoc Committee on Health Research Relating to Future Intervention Options of the World Health Organization has set forth a methodology to determine and specify "best buys" in research for both communicable and non-communicable diseases. *Investing in Health Research and Development: Report of the Ad Hoc Committee on Health Research Relating to Future Intervention Options* (Geneva: World Health Organization, 1996).

6. A specific analysis of the process of changing from geography-based management organization is provided by Thomas Malnight in his description of Citibank's reformulation of its European corporate banking operations between 1979 and 1994. Thomas W. Malnight, "The Transition from Decentralized to Network-Based MNC Structures: An Evolutionary Perspective," *Journal of International Business Studies* First Quarter 1996, 43–65.

Appendix One: Global Health Institutions and Resource Patterns

W̲hat is the current constellation of international health institutions, and how are resources spent? There are a great number of government, non-government, and private actors on the international health stage. There is no regular, formal inventory, current or historical, of these institutions or programs, their substance or functional focus, their total resources, or their resource allocation patterns. Hence, drawing a full and complete picture over time of the institutional patterns with regard to international health allocations would be a herculean task.

One can, however, take a measure of the most significant organizations on the international health stage, and at least have a core picture of current program portfolios and resource levels. We assume that the programs of other organizations represents a pattern similar to the "market leaders."

The World Health Organization

The World Health Organization (WHO) is certainly a key leader. WHO was created at the conclusion of World War II to provide a mechanism for international discussion and action on health care issues. Headquartered in Geneva, WHO has 6 regional offices and programs throughout the world. In the period 1994–95, the WHO budget was $1.84 billion, with 46% derived from the "general budget" of contributions by all nations, and 54% derived from specific contributions targeted at special programs and trust funds, such as tropical diseases research or AIDS, and/or in transfers to WHO from other UN agencies for program implementation.

Appendix 2 provides a summary of the level of variety in WHO's program portfolio and resource allocation among those programs.

WHO programs are divided into four categories, which encompass a total of 14 program areas and 96 activity sub-categories (exclusive of internal management activities). In recent years, WHO resources have been held at zero growth levels, and in the last two biennial budgets have begun to decline in real terms. Table 1 provides a summary of resource allocation trends over the last two biennium budget periods. (Appendix 2 provides programmatic definitions for each of the 6 categories summarized here.)

These totals combine regular budget funds, and extra-budgetary funds, the latter provided by a limited number of nations (the majority of funds come from 10 nations) in support of specified program initiatives. Extra-budgetary funds now make up over 50% of WHO's total budget, compared to 20% in the 1970s.

Geographic distribution of WHO resources is summarized in Table 2.

WHO allocates its program resources at four levels: activities within countries, activities between countries, activities at the level

Table 1. WHO Budget Trends.

Program Category	Budget($ million)		Budget (% total)	
	1994–95	1996–97	1994–95	1996–97
Governing Bodies	15.7	21.9	0.8	1.2
Health Policy Management	487.8	472.1	23.9	26.6
Health Services Development	271.0	233.5	13.7	12.2
Promotion/Protection of Health	355.7	325.0	18.0	17.7
Integrated Disease Control	646.7	590.4	32.8	32.2
Administrative Services	194.4	193.5	9.7	10.5
Total	1971.3	1836.4		

of the geographic region, and activities which combine regions or which are conducted/managed at the global, headquarters level. Table 3 provides a resource distribution summary for this allocation pattern.

Yet, as critical as WHO is, it now represents only about 10% of the resources currently flowing for health at the international level. A rich fabric of institutions has evolved over the last 40 years that has both expanded and deepened international health capacity.

The World Bank

To nearly $1 billion per year in international health resources of WHO, one needs to add the resources of the World Bank. Lending for "population, health and nutrition" (PHN) activities has become a significant part of the Bank's portfolio. Indeed, PHN lending has risen from 2.5% of Bank lending in 1989 to an estimated 10.9% in 1996. Between 1989 and 1996, the Bank has committed to 10.2 billion in PHN lending, or about $1.2 billion per year. However, real lending per year is extremely variable as noted in Graph 1. Appendix 3 provides detailed breakdowns of 1989–1996 PHN lending.

Table 2. Regional Resource Allocation.

	Budget ($ million)		Budget as % of total	
	1994–95	1996–97	1994–95	1996–97
Africa	245.9	211.2	12.5	11.5
The Americas	404.9	312.4	20.5	17.0
South-East Asia	233.6	174.3	11.9	9.5
Europe	94.0	76.4	4.8	4.2
Eastern Mediterranean	102.6	95.4	5.2	5.2
Western Pacific	105.9	90.2	5.4	4.9
Global & Interregional	784.5	876.4	39.8	47.8
Total	1971.4	1836.3		

Table 3. Resource Allocation By Operational Level

	Budget ($ million)		Budget as % Total	
	1994–95	1996–97	1994–95	1996–97
Country	613.0	451.2	31.1	24.6
Intercountry	420.3	361.3	21.3	19.7
Regional Offices	153.5	147.5	7.8	8.0
Global & Interregional	784.5	876.4	39.8	47.8
Total	1971.3	1836.4		

The most informative data is disbursement rates on the lending, since this shows the flow of resource allocation rather than the annual stock of lending. Graph 2 illustrates disbursement patterns 1986–96. Total disbursements for health, population, and nutrition now total about a billion dollars a year, and have shown a significant increase since 1992.

Graph 1. World Bank PHN Lending 1989–1996.

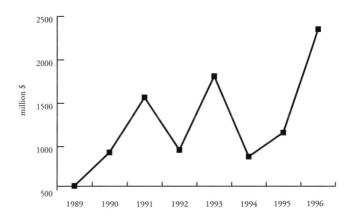

Graph 2. IBRD/IDA Disbursements
Health, Nutrition & Population Sector.

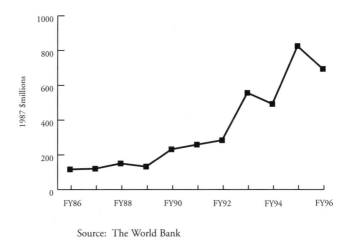

Source: The World Bank

In terms of geographic distribution the nations of South Asia and Latin American have received about half of the Bank's PHN lending total, with the remainder spread in the Bank's other four lending regions. Table 4 summarizes these trends.

The PHN portfolio rarely exceeds 5–6% of the Bank's total lending in a region. The regular exception is South Asia, where the PHN level was below 6% of portfolio in only two of the 8 years analyzed. Sporadic exceptions are also seen in Africa, Europe, and Latin America.

The Bank lends both "soft loan" money through its IDA window and more structured money thought its IBRD arm. One surrogate for the link between the lending portfolio and the income levels of countries receiving loans is to assess the distribution of the lending as between these two mechanisms. In the late 1980s and early 1990s, PHN lending stood in contrast to overall Bank lending patterns. PHN lending was about evenly split between IDA and IBRD windows, while overall bank lending was about 25% IDA and 75% IBRD. In the 1992–1995 period, the PHN portfo-

Table 4. PHN Lending by Region ($ million)

	1989–96 Total PHN loan commitments	% Total PHN lending	Total 1989–96 PHN disbursements
Africa	1,610.1	15.7	751
East Asia/Pacific	1,357.8	13.3	707
South Asia	2,777.8	27.2	1,431
Europe & Central Asia	1,016.8	9.9	168
Latin Am. & Caribbean	2,757.0	27.0	950
Middle East & N. Africa	703.4	6.9	321
Total	10,222.9	100.0	4,328

lio became heavily IDA dominated, with a record 68% of PHN lending being in the form of IDA soft loans in 1992. The overall Bank pattern began to shift a bit more toward the IDA window, but still only about 30% was soft loan and 70% IBRD lending. In 1996, the PHN patterns shifted dramatically, with 36% of lending in soft loans, and 63% in harder IDA loans, a near mirror image of the Bank's overall lending distribution. Appendix 3 summarizes lending patterns by region and in terms of the IDA/IBRD split for the period 1989–96.

It is extremely difficult to determine World Bank PHN lending by disease or health condition problem or by function. Projects tend to cover a variety of functions (planning, training, management, procurement) and address a range of health problems.

In addition, of course, the Bank is a leader in many multilateral health efforts, including the Tropical Disease Control Programme and the Population Programme of WHO. The Bank, however, acts as a technical partner and co-sponsor but not as a major source of financing. The Bank also engages in and commissions research in a variety of international health planning and disease control areas.

Regional Banks

The regional banks are also significant institutional actors. The Inter-American Development Bank lends about $1.5 billion per year for health and education. The Asian Development Bank commits about $50 million per year to health projects. The European Bank for Reconstruction and Development has also lent for health system reform in the newly democratic nations of Central and Eastern Europe, but at lesser levels. The African Development Bank (ABD) allocated about $34 million to health loans in 1994, or about 2% of total lending. Between 1967 and 1995, health accounted for 8% of total ABD lending.

UNICEF

UNICEF's 1995 budget totaled $1.023 billion, a 2.6% increase over 1993 which is effectively a no-growth budget in real terms. UNICEF's total income is derived from three sources. General resources (54% of the total) are derived from annual contributions of 104 governments, contributions of National Committees, and sales of greeting cards. Emergency funds (16% of the total) are targeted contributions for emergency response. Supplementary funds (30% of the total) are from governments and intergovernmental organizations to support projects for which general resources are insufficient or for sudden relief or rehabilitation incidents.

About 90% of UNICEF's budget is allocated to programs. Functionally, 36% is used for procurement of supplies (medicines and vaccines), 52% for grants and technical assistance for field projects providing health, education, and nutrition, and 12% for administrative and other support to programs. Table 5 provides

Table 5. UNICEF Expenditures ($ million)

Budget Category	1993 Amount (%)	1994 Amount (%)	1995 Amount (%)
Funders			
Special Assistance	359 (40%)	334 (37%)	330 (36%)
Cash & Other Assistance	445 (50%)	467 (52%)	474 (52%)
Programme Support Services	93 (10%)	99 (11%)	109 (12%)
Subtotal	897	990	913
Other			
Administrative Services	87	91	99
Write-offs & Other Charges	13	8	11
Total	997	999	1,023

program comparisons from 1993 through 1995.

In the last five years, UNICEF has experienced a fairly dramatic shift in the programmatic uses of its resource. As a percentage of total program resources, child health, water, and nutrition programs have receded, and emergency relief, education, community development, and planning/advocacy have grown. Table 6 provides comparisons of programmatic resource allocation as a percentage of program expenditures.

Table 6. UNICEF Program Expenditure Comparisons, 1991 and 1995.

Category	% of Program Budget	
	1991	1995
Child Health	34.2	26.4
Water Supply & Sanitation	12.4	8.8
Child Nutrition	5.2	3.9
Community Devt/Women's Programs	6.6	8.1
Education & Early Child Devt	8.1	10.6
Planning, Advocacy, and Prog Support	14.7	17.0
Emergency Relief	18.8	25.3

UNICEF has country program agreements with 145 countries. Geographically, UNICEF's 1996–2000 budget projections show a particular emphasis on Africa and Asia. Table 7 provides budget projections by region.

Bilateral Programs—The United States

Among bilaterals, the programs of the United States remain significant on the international scene. U.S. government allocations to

Table 7. Multi-year Regional Distribution of General Resources: UNICEF Programs

Sub-Saharan Africa (ESARO and WCARO)	$608,760,000
Latin America and the Caribbean (TARCO)	$116,520,000
Asia (EAPRO and ROSA)	$583,740,000
Middle East and North Africa (MENARO)	$89,761,000
Central and Eastern Europe, the Commonwealth of Independent States (CEE/CIS) and the Baltic States	$60,900,000
Total	$1,459,681,000

international health are spread across a great number of agencies and programs. The last detailed inventory of these resources was completed in 1979; more recent integrated data are not available. At that time, the total U.S. government commitment to international health was $563 million in fiscal 1980, with 4% coming from defense agencies, 76% from the Agency for International Development, and the remaining 20% from other civilian agencies.

The Agency for International Development remains a key actor in the U.S. government's international health pantheon. AID spends about a billion dollars a year on health, population and nutrition programs. These expenditures include both bilateral projects and U.S. contributions to several programs of WHO and UNICEF, either through grant transfers or through joint research or project implementation. AID expenditures also include resource transfers to other U.S. agencies (e.g., the Centers for Disease Control or the Office of International Health and Refugee Affairs of the Department of Health and Human Services) for international health programs.

Table 8 provides a summary of AID population, health, and

Table 8. AID Population/health Funding ($ millions)

	FY 1994	FY 1995	FY 1996
Population	480	511	432
Child Survival	243	272	300.8
HIV/AIDS	112.8	118.8	114.2
Other Health[1]	160[2]	149.8	127.4
Total	995.8	1050.6	973.4

[1]About one half of "other" funding is ESF for Egypt and the New Independent States.
[2]estimate.

nutrition appropriations over the last three fiscal years. As with World Bank data, these figures represent total annual appropriation not annual resource flows. The total annual disbursement on health at AID is not known, although a new data tracking system is now being developed toward that end.

There is, therefore, some double counting even at this sketchy profiling of international health resources. An AID outflow may be a WHO inflow. These amounts are small, however, relative to total resource aggregates. About one percent of the AID population/health/nutrition total expenditure represents income to other U.S. agencies. In Fiscal 1995, all U.S. Government activities in collaboration with WHO totaled only about $6.6 million.

Two other U.S. civilian agencies are also key, the National Institutes of Health and the Centers for Disease Control.

The United States participates in 28 bilateral umbrella S&T agreements, and over 800 agency-to-agency agreements and memoranda of understanding. Within those 800 agreements, The Department of State Title V report indicates that 59 relate to biomedical science, and 4 relate to health and medical sciences. The NIH participates in 55 of these agreements.

In fiscal 1995, NIH international programs totaled $186 million. Approximately 43% of this total was allocated to foreign research grants and foreign components of U.S. grants, and 50% to awards for scientific exchanges. The remainder of the funding is allocated to various other programs and to international travel.

Less than one percent of the research and research training awards funding was a transfer from NIH to international organizations. Hence, bilateral research monies, although international in nature, represent very separate flows on the international health institution scene. It is interesting to look at the level of U.S. research to international health problems relative to the resources of international organizations targeted at these issues. As illustrated in Table 9, U.S. resource commitments for research are much larg-

Table 9. Illustrative Disease Research ($million): Resource Commitments

	FY1995 USG Research Grants**	World Health Organization* Research	World Health Organization* Control
Leprosy	8.86	2.35	4.17
Malaria	28.50	8.00	11.65
Diarrheal Diseases	22.80	3.70	10.00
Tuberculosis	80.80	n/a	10.10
Polio	13.50	n/a	n/a

*WHO data are available for the 1994–95 biennium not on an annual basis; hence these data are one half the biennium number.

** these are conservative estimates as they total only those grants for which the Critical Technologies Institute's Radius data base contains average annual outlay data, not those research grants registered but without annual budget estimates.

er than the equivalent programs of, for example, WHO and, indeed, much larger than the total of R&D plus programmatic budgets for control. The importance of careful targeting of the "international dollar" in these areas is thus illustrated.

The Centers for Disease Control operates global programs which total $14 million in project resources distributed across the organization. The CDC efforts are carried out in collaboration with NIH institutes, other international organizations (e.g., WHO and the World Bank), and foreign epidemiological agencies. CDC allocates about one percent of its budget and four percent of its staff to global activities.

If the relative weight of international health resources among U.S. agencies remains approximately the same as in the early 1980s, then U.S. bilateral resources in international health probably total about $1.3 billion. If the 1979–80 proportions hold approximately true in the mid-1990s, moreover, about 80% of this total is for program efforts in foreign countries or regions, and the remaining 20% is focused at the global level. ■

Appendix Two: WHO Programme Budget, 1994-1997

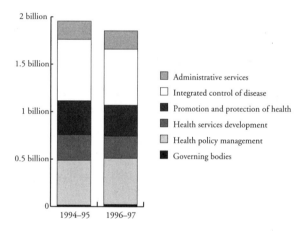

Table 1. WHO Programme Budget, 1994–97 (U.S. dollars)

	1994–95	1996–97
Governing bodies	15,691,300	21,912,800
Health policy management	472,061,000	487,827.600
Health services development	270,976,400	233,462,000
Promotion and protection of health	355,706,700	325,029,000
Integrated control of disease	646,676,600	590,405,200
Administrative services	194,420,000	193,504,000
Total	1,955,532,000	1,852,140,600

1.   **Governing bodies**

       1994–95   15,691,300

       1996–97   21,912,800

2.   **Health policy management**

       1994–95   472,061,000

       1996-1997   487,827,600

2.1   *General programme development and management*

       1994–95   108,284,800

       1996–97   104,459,000

Activity categories for 1996–97

- Program policy
- Management (planning, monitoring, evaluation)
- Information
- Staff and management development
- Internal audit oversight

2.2   *Public policy and health*

       1994–95   36,299,400

       1996–97   45,862,100

Activity categories for 1996-1997

- Health in socioeconomic development
- Women in health and development
- Health legislation, human rights and ethics
- Research policy and strategy coordination

2.3   *National health policies and programmes
development and management*

       1994–95   222,128,100

1996–97   225,174,900

Activity categories for 1996-1997
- International cooperation for health
- Planning, management and coordination of emergency and humanitarian action
- Emergency preparedness
- Relief and rehabilitation
- Protection from violence
- Procurement and supply chain management

2.4   *Biomedical and health information trends*
1994–95   121,115,300
1996–97   96,564,500

Activity categories for 1996–97
- Global health situation analysis and projection
- World health reporting
- Strengthening country health information
- Coordination and cooperation in epidemiological and statistical activities and trend assessment
- Language services
- Publishing activities
- Dissemination of publications
- Library and information services

3.   **Health services development**
1994–95   270,976,400
1996–97   233,462,000

3.1   *Organization and management of health systems based on*

primary health care
1994–95   109,338,600
1996–97   92,023,700

Activity categories for 1996–97
- Health systems research
- Policy reform and restructuring of health systems
- Health financing and economic: option appraisal, systems and skills development
- District, local and community health action
- Hospitals and health centres: performance and quality assurance (global and all regions)
- Health systems response to rapid urbanization (global and all regions)
- Information and telematics support to health care

3.2   *Human resources for health*
1994-1995   103,397,700
1996-1997   76,874,200

Activity categories for 1996-1997
- Multidisciplinary approaches
- Policy and planning
- Methodology development
- Management of health personnel
- National management information systems
- Development of educational programmes
- Evaluation and quality assessment
- Changes in health personnel education and practice

- Nursing policy
- Nursing/midwifery practice
- Theory and practice of new public health
- Public health training
- Continuing education
- Health learning materials
- Fellowships

3.3   *Essential drugs*

1994–95   34,277,300

1996–97   37,769,00

Activity categories for 1996-1997
- Country support
- Development work
- Operational research
- Management and advocacy
- Procurement of drugs and biologicals

3.4   *Quality of care and health technology*

1994–95   23,962,800

1996–97   26,795,000

Activity categories for 1996-1997
- Quality of care and health technology assessment
- Clinical technology
- Blood safety
- Health laboratory technology
- Radiation medicine
- Drugs and biologicals, quality safety and efficacy
- Traditional medicine

4.   **Promotion and protection of health**

        1994–95   355,706,700

        1996–97   325,029,000

4.1   *Reproductive, family and community health and population issues*

        1994–95   130,753,200

        1996–97   112,087,100

Activity categories for 1996-1997

- Family health
- Child health and development
- Family planning and population
- Maternal health and safe motherhood
- Adolescent health
- Women's health
- Human reproduction
- Occupational health
- Aging and health

4.2   *Healthy behavior and mental health*

        1994–95   53,414,800

        1996–97   48,215,400

Activity categories for 1996-1997

- Strengthening community-based interventions
- National capacity-building
- Settings for health promotion
- Special population groups
- Assurance of quality care and equitable access to treatment
- Training

- Promotion of research and transfer of know-how
- Health communications and public relations
- Advocacy, networking and collaborations

4.3　　*Nutrition, food security and safety*
　　　1994–95　　48,628,800
　　　1996–97　　52,676,600

Activity categories for 1996-1997
- Prevention of malnutrition and promotion of safe food and healthy nutrition
- Monitoring and surveillance
- Emergencies
- Research, training and development
- Partnerships

4.4　　*Environmental health*
　　　1994–95　　122,909,900
　　　1996–97　　112,049,400

Activity categories for 1996-1997
- Water supply and sanitation
- Assessment of environmental health risk and information for management
- Chemical safety

5.　　Integrated control of disease
　　　1994–95　　646,676,600
　　　1996–97　　590,405,200

5.1　　*Eradication/elimination of specific communicable diseases*
　　　1994–95　　25,871,500

1996–97   37,566,500

Activity categories for 1996-1997
- Dracunculiasis
- Leprosy
- Poliomyelitis
- Neonatal tetanus

5.2   *Control of other communicable diseases*
1994–95   540,902,200
1996–97   476,473,000

Activity categories for 1996-1997
- Vaccine-preventable diseases and immunization
- Epidemiological surveillance and statistical services
- Acute respiratory infections
- Diarrhoeal diseases
- AIDS/sexually transmitted diseases
- Tuberculosis
- Malaria and other tropical diseases
- Vector control
- Tropical diseases research
- Other communicable diseases, including zoonoses and surveillance of emerging diseases and antibiotic resistance
- Blindness and deafness

5.3   *Control of noncommunicable diseases*
1994–95   79,902,900
1996–97   76,365,700

Activity categories for 1996-1997

- Cardiovascular diseases
- Cancer
- Other noncommunicable diseases

6.   Administrative services
1994–95   194,420,000
1996–97   193,504,000

6.1   Personnel
1994–95   22,927,600
1996–97   23,442,500

6.2   General administration
1994–95   127,411,200
1996–97   124,293,600

6.3   Budget and finance
1994–95   44,081,200
1996–97   45,768,500

Bibliography

African Development Bank. *Compendium of Statistics.* Abidjan: African Development Bank, 1996.

Bleecker, Samuel E. "The Virtual Organization." *The Futurist.* (March–April, 1994).

Cahners Research. *Basic Research.* Des Plaines, IL: Cahners Research, 1997.

Cohen, David R. "Messages from Mid Glamorgan: a multi-programme experiment with marginal analysis." *Health Policy* 33 (1995), 147–155.

Commission on Health Research for Development. "Health Research: Essential Link to Equity in Development." New York: Oxford University Press, 1990.

Donaldson, Cam. "Economics, public health and health care purchasing: reinventing the wheel?" Health Policy 33 (1995), 79–90.

"Fast-tracking epidemic information," *The Lancet,* (December 14, 1996), 1599.

Frankel, David H. "Americas act to combat dengue outbreaks." *The Lancet,* October 7, 1995, 957.

Frenk, Julio, *et al.* "Health Transition in Middle-income Countries: New Challenges for Health Care." *Health Policy Planning* 4 (1), 1989, 29–39.

Frenk, Julio, and Fernando Chacon. "International Health in Transition." *Asis-Pacific Journal of Public Health* 5, 1991.

Frenk, Julio, "Comprehensive Policy Analysis for Health System Reform," in Berman, Peter (eds) *Health Sector Reform in Developing Countries: Making Health Development Sustainable.* Harvard Series on Population and International Health, (Cambridge: Harvard School of Public Health, 1995).

Julio Frenk, *et al.* "The New World Order and International Health." *British Medical Journal* 314 (May 10, 1997).

Gwynne, Peter. "Managing 'Multidomestic' R&D at ABB." *Research–Technology Management.* Industrial Research Institute, 1995.

Hane, Gerald, and Gary Hufbauer. *The Globalization of Industrial Research and Development, Seesionb on R&D Globalization and Trade Policy.* Council on Foreign Relations Study Group, Session on R&D Globalization and Trade Policy (13 March 1997).

"International Health Report of the International Health Subcommittee, Development Coordination Committee." Volume II, Detailed Tables. December, 1980. [place of publication?]

Interview with Percy Barnevik. *Harvard Business Review.* "The Logic of Global Business: An Interview with ABB's Percy Barnevik." Cambridge: Harvard Business School, 1991.

Jennett, Penny, "The Canadian Network for Technology-Based Learning." *New Currents* (January 1995).

Kimball, Ann Marie, and Myo Thant. "A role for businesses in HIV prevention in Asia." *The Lancet,* (June 15, 1996), 1670-1672,

Knutsson, Karl Eric, Goran Tomson, and Karl-Olaf Wathne, *Alliance for Health Policy/Systems Research: Report and Proposals from an International Consultation.* Swedish International Development Cooperation Agency and Royal Ministry of Foreign Affairs, Norway (June 6, 1997).

Madden, Lynne, Ruth Hussey, Gavin Mooney, and Elaine Church. "Public Health and Economics in Tandem: Programme Budgeting, Marginal Analysis and Priority-setting in Practice." *Health Policy* 33 (1995), 161–168.

Malnight, Thomas W. "The Transition from Decentralized to Network-Based MNC Structures: An Evolutionary Perspective." *Journal of International Business Studies.* First Quarter, 1996.

Ministry of Foreign Affairs, Stockholm, Sweden. *Tomorrow's Global Health Organization: Ideas and Options.* Stockholm: Norstedts Tryckeri, 1996.

Moffet, John. "Environmental Priority-setting Based on Comparative Risk and Public Input." *Environmental Priority Setting.* From the Canadian Public Administration, Vol.39, No.3, Fall [year?].

Morse, Stephen S., Barbara Hatch Rosenberg, and Jack Woodall. "ProMED Global Monitoring of Emerging Diseases: Design for a Demonstration Program." *Health Policy* 38 (1996) 135–153. Washington, D.C., Federation of American Scientists, ProMED Steering Committee.

Murray, Christopher J. L., and Alan D. Lopez. *The Global Burden of Disease.* Washington, D.C.: World Health Organization, 1996.

National Institutes of Health. *Briefing Book for Foreign Visitors to the National Institutes of Health.* Washington, D.C., 1996.

National Institutes of Health, Division of Microbiology and Infectious Diseases. *The Jordan Report.* Bethesda, MD: National Institute of Allergy and Infectious Disease, National Institutes of Health, 1997.

Nau, Jean-Yves. "Preventing Spread of BSE." *The Lancet,* (September 17, 1994), 808.

Ponka, Antti, *et al.* "Salmonella in Alfalfa Sprouts." *The Lancet,* (February 18, 1995), 462–463,

Press Release from the Ad Hoc Committee on Health Research Relating to Future Intervention Options. "Non-Communicable Diseases to Become Leading Global Cause of Death: Largest Increase Comes in Developing World." Washington, D.C., Sunday, September 15, 1996 at 4:00 PM.

Rayport, Jeffrey F. "Exploiting the Virtual Value Chain." *Harvard Business Review.* Cambridge: Harvard Business School, 1995.

Rosencrance, Richard. "The Rise of the Virtual State." *Foreign Affairs.* 75 (July–August, 1996).

Rogers, Arthur, "European Union's Role in Public Health Unsettled." *The Lancet,* (April 20, 1996), 1107.

Sepulveda, Jaime, *et al.* "Key Issues in Public Health Surveillance for the 1990's." *International Symposium on Public Health Surveillance* (April, 1992).

Shinnar, Shinnar, Paul Citron, and David Hodges. *The Globalization of Industrial Research and Development.* Council on Foreign Relations Study Group, Session on Industries and Technology, (12 December 1996).

Smith, Raymond W. "Bell Atlantic's Virtual Work Force." *The Futurist.* (March–April, 1994) [page refs].

Studt, Tim, editor. "Basic Research White Paper: Defining our Path to the Future." *Research and Development,* n.d.

The International Bank for Reconstruction and Development, World Bank. *World Development Report 1993: Investing in Health.* New York: Oxford University Press, 1993.

The Rockefeller Foundation, Social Science Research Council, "Enhancing the Performance of International Health Institutions." Pocantico Retreat. Harvard School of Public Health, February 1–3, 1996.

United Nations Children's Fund. "1996 Unicef Annual Report." New York, 1996.

United Nations Children's Fund. "The State of the World's Children 1996." Oxford: Oxford University Press, 1996.

United Nations Children's Fund. "The Progress of Nations: The Nations of the World Ranked According to their Achievements in Child Health, Nutrition, Education, Family Planning, and Progress for Women." Benson, Wallingford: P&LA, 1996.

Williams, Alan. "Qalys and Ethics: A Health Economist's Perspective." *Soc. Sci. Med.* 43 (12). Great Britain: Elsevier Science Ltd, 1996.

Wilson, Mary E., Richard Levins, and Andrew Spielman. *Disease in Evolution: Global Changes and Emergence of Infectious Diseases.* Annals of the New York Academy of Sciences, 740. New York: New York Academy of Sciences, 1994.

World Health Organization. "Tropical Disease Research: Progress 1975–94, Highlights 1993–94." *Twelth Programme Report of the UNDP/World Bank/WHO Special Programme for Research and Training in Tropical Diseases (TDR).* Geneva, 1995.

World Health Organization. "Programme Budget for the financial period 1996–1997." Geneva, 1995.

World Health Organization. "Investing in Health Research and Development: Report of the Ad Hoc Committee on Health Research Relating to Future Intervention Options." Geneva, 1996.

World Health Organization. *The World Health Report 1997.* Geneva, 1997.

Wu, Trong-Neng *et al.* "Hantavirus Infection in Taiwan", *The Lancet,* (March 16, 1996), 770–771.

2

Research Priorities

Globally, the research community spends in excess of $60 billion on biomedical research each year, of which about 50% is invested by the private sector. Setting priorities for this research is complex, and fraught with cross-purposes. There is never (perhaps will never be) enough money to explore every question, pursue every opportunity, solve every problem, save every life. Yet, research advance requires money (often lots of it) and an acceptance that tangible evidence of the utility of the investment in research may lie years in the future. The dynamics and flows of the marketplace do not always fit well with the capacities of academic institutions. Yet, the partnership of industrial and university capacity is essential for research to reach people in need. Public policy, and public budgets, often are produced neither by the marketplace nor by academic excellence, but by politics driven by gusts of public opinion. Yet forging dependable public policy out of the vagaries of politics is essential to building public support for health research in pluralistic democracies with a myriad of other societal needs.

What Could Be Simpler Than Childhood Vaccination?: A Parable for Health Care Reform

Based on remarks
delivered at the
Science Policy
Association of the
New York Academy of
Sciences on October
11, 1994.

BARRY R. BLOOM
Albert Einstein College of Medicine

I mmunization is recognized worldwide as the most cost-effective medical intervention for preventing death and disease. In 1992, one major vaccine cost as little as 55¢ per pediatric dose; in 1993 that figure was as low as 35¢ (down from $15 per pediatric dose in 1986). Since 1974, the World Health Organization (WHO) and the UN Children's Fund (UNICEF) have promoted early-childhood vaccination programs around the world. Indeed, immunization rates in the Third World reached between 80 and 85 percent in 1994. In contrast, U.S. vaccination programs remain a frustrating public health problem. An examination of how the nation ended up with such a poor immunization record, and a discussion of what can be done about it can provide important lessons for what might work, and what certainly won't, in the context of national health care reform.

U.S. Childhood Vaccination Rates: A Tragic Unknown

> One of the tragedies of this very sophisticated country is that we have virtually no comprehensive data about what percentage of our children are immunized.

In the United States, early-childhood vaccination rates are often lower than those of the world's poorest countries. A 1992 report by the Centers for Disease Control estimated that in most cities surveyed, fewer than 50 percent of children were fully immunized by age two. The problem, moreover, is not found merely in pockets of poverty or among children far removed from health care facilities. A survey of the employees of the health care corporate giant, Johnson & Johnson, estimated that only 45 percent of employee's children were fully vaccinated by age two, and only 55 percent by age six.

Yet, even such shocking figures are only rough indicators. One of the tragedies of this sophisticated country is that we have virtually no comprehensive information about what percentage of our children are immunized.

A Public Sector Tower of Babel

This ignorance is due to the fragmentation of agencies and the lack of unitary goals that characterize the nation's vaccination system. This potpourri of approaches yields little in the way of coordinated data collection. In the state of Massachusetts, for example, 85 percent of children are vaccinated by private physicians, with the state's role limited to ensuring availability of vaccines. In most southern states, on the other hand, 85 percent of children are vaccinated at public health clinics. And everything else in between is everything else in between. Thus, compiling and analyzing comprehensive data is difficult. What is needed is a registry and reminder system that tracks children's immunization status, whoever is the health provider examining the child.

Currently, some twenty-four federal executive branch agencies, several Congressional committees, fifty state programs, and innumerable health care insurance plans play some role in the nation's immunization effort. There is no central coordinating function. All of these institutional actors compete for resources and for authority in what is an enormously complex undertaking.

The Private Sector: An Indispensable Partner Faces Major Problems

Private industry is a central partner in the immunization equation. Private enterprise undertakes the applied research, develops vaccines, manufactures the product, and stores the supplies that the nation needs. For the private sector, however, shifting government roles and tidal changes in financing strategies lead to critical market uncertainties and hence dampen corporate interest in the vaccine enterprise.

And that enterprise is expensive, lengthy, and fraught with risk. Bringing a new vaccine from the laboratory to the point of licensure now requires $200 million and eleven years of effort. Given the cost and risk involved, corporate effort requires the assurance of reasonable prices and large markets.

U.S. corporations, however, have faced two severe challenges to these prerequisites. The industry has been subject to government threats of domestic price controls, which would seriously undermine its financial viability. Moreover, although expanding sales to international markets would enable U.S. companies to reduce vaccine unit costs, this alternative has been simultaneously undercut by government policy. Nearly 40 percent of the world's vaccines are purchased by UNICEF for distribution to the world's poor. To compete for UNICEF procurements, U.S. companies would need to offer vaccines at lower prices than those charged in the U.S. The U.S. Congress has been critical of two-tiered pricing, despite the fact that being able to compete for such procurement would enable companies to expand production, achieve economies of scale, and hence lower prices in the U.S.

Such fundamental conflict within public policy serves as a major source of disincentives for private sector investment and innovation in the U.S. health care system, discouraging research and development for vaccines for both new and traditional diseases. Ironically, the negative policy environment coincides with a scientific environment whose potential for important vaccine innovations has never been more promising.

A Policy That Works: Vaccine Injury Compensation

In 1993, there were fourteen U.S. vaccine manufacturers. Today there are four. One of the reasons for this decline is the major threat of liability suits. As with all other medical interventions, there is always the small risk of unforeseeable adverse effects of vaccines in a small number of children. There is at least one precedent, however, that provides an example of how government policy can both protect the public good and enhance the competitiveness of private initiative.

The Omnibus Health Act of 1993 established a liability system for childhood vaccines that significantly removed the excess costs of litigation from the industry. In essence, Title XXI of the Act is the first effort since worker's compensation to provide a system of no-fault health insurance. It represents a profound change in the way the "business" of medicine is carried out.

Title XXI includes the National Vaccine Injury Compensation Program. An excise tax, paid by the consumer or insurer, is levied on every vaccine dose. The revenues from this tax are placed in a public trust fund which is used to settle claims brought by consumers (and found by the National Academy of Sciences to have merit), without the necessity of legal suits. Experience to date indicates that, for the 4 million children vaccinated in the U.S. every year, only 150 cases are brought to the compensation process, and only 4 percent of these are deemed valid by the review process.

> Currently, there are some twenty-four federal executive branch agencies, several Congressional committees, fifty state programs, and innumerable health care plans that play some role in the nation's immunization effort.

A system is in place, therefore, that both motivates private industry to remain involved in vaccine development and production, and satisfies the reality of medical risk. Such a straightforward mechanism provides an important example of how private roles and the public good can be made compatible within the U.S. health care system.

What Is Needed: A National Vaccine Authority

However, on a broader scale the U.S. immunization programs require a mechanism for coordinating vaccine research, safety, quality, and provision among public agencies and between private industry and the government. They require a "National Vaccine Authority." Such an Authority, advocated by the Institute of Medicine in its 1993 report on the nation's immunization programs, would not aim to control immunization programs nor to supplant the role of private industry either in research and development or in production. Rather, the Authority would carry out the mandates of the 1986 Public Health Service Act, which created a

> **It now takes $200 million and eleven years to get a new vaccine from the laboratory to licensure. To recoup such investments will require a climate of trust between private manufacturers and the government.**

National Vaccine Program to achieve optimal disease prevention through improved immunization. The legislation also created the National Vaccine Advisory Committee, which is the only forum currently available in which all parties, private and public, can meet to discuss and resolve national vaccination issues.

The Committee and Program, however, are currently victims of the budget axe. Of thirty-five professional positions allocated to the Committee to provide technical backstopping to the coordination process, thirty staff were returned to their original agencies as of October 1, 1994. In essence, then, the federal government's right hand, via the legislatively authorized Committee, moved toward coordination, while the government's left hand, via the budget process, is simultaneously pushing to reduce coordination and public participation.

In this push-pull setting, what is clearing lacking is leadership. The National Vaccine Advisory Commission has the mandate to exercise that leadership. It is unfortunate that it has not been enabled to do so.

How should that be accomplished? First, the Commission requires a small, competent staff to support policy analysis and program innovations. This must be a continuous staffing function. The Secretary of Health and Human Services has the legislative authority to spend 1 percent of Public Health Service Funds on evaluation of vaccine programs. This funding could be used to support Commission work.

Second, Commission leadership will require the ability to respond to national needs or emergencies. The National Vaccine Injury

Compensation Program (NVICP) has a large trust fund, a portion of which could be used flexibly to respond to needs on an immediate basis.

> In the push-pull world of national vaccination policy, what is clearly lacking is leadership.

Third, the Commission would need to develop and maintain a national immunization registry to provide a systematic basis for ensuring that all children are immunized. The groundwork for such a registry has already been laid by the National Vaccine Plan. What is needed is leadership toward implementation.

What Is Not Needed: More Legislation and Budget Inflation

Such solutions for rationalizing immunization programs in the U.S. rest comfortably within existing patterns of legislative authorizations. Vision and leadership, not more legislation, are needed to implement programs using those authorizations.

Similarly, the authority to allocate resources to immunization reforms is already present within the U.S. Public Health Service. Acting on a clear vision does not necessarily imply ever-widening resources. It does require placing priority on immunization programs when existing resources are allocated.

On the other hand, it is clear that the leadership to implement immunization reform will not be possible if cuts in existing resources remove the ability to act on this national priority. Drastic budget reductions will put fundamental aspects of the nation's health at risk. Many people desire smaller government, but public health is not the place to save money.

Lessons for Future Health Care Reform

The opportunities for action to dramatically improve the U.S. immunization system provide several insights that might equally

guide future approaches to health
care reform overall.

First, the vaccination system is a
complex enterprise that requires
long-term commitments from both
industry and public agencies, as well
as knowledge and participation of

> **Many people desire small-
> er government, but public
> health and disease
> prevention are not the
> place to save money.**

the citizenry. Barriers and disincentives in any one part of the system
affect the entire enterprise. Addressing the problems and enlisting
the cooperation of all parts of the system is a prerequisite to success.

Second, private industry roles are critical. Risk-taking in the laboratory and competition in the marketplace will lead to more and
better vaccines. Public private collaboration, not confrontation,
must underpin efforts to improve vaccination programs.
Mechanisms are necessary for ensuring that both the public and
the private sectors are involved in program decision making from
the beginning of reform efforts and continually thereafter.

Third, a stable and predictable public policy environment is also
important. Wide swings of the policy pendulum remove the basis
for long-term trust among the various groups working within the
vaccination enterprise. The same has been true and will be true of
efforts at national health care reform.

Fourth, the immunization system rests on a pillar of public
understanding and compliance. Policy decisions and new systems
to implement those decisions must be transparent and must
involve broad public participation. Effective reform will require
deep public understanding and broad public cooperation.

Finally, new initiatives and new solutions do not necessarily
require budget-breaking financial appropriations. They do require
maximizing existing health care institutions, choosing clear national priorities, and allocating resources to those priorities. ∎

Academic Health Centers Face the Future: Reform, Risk, and Restructuring

Based on remarks delivered at the Science Policy Association of the New York Academy of Sciences on November 17, 1994. In 1995, Dr. Richardson became president and chief executive officer of the W. K. Kellogg Foundation in Michigan.

WILLIAM C. RICHARDSON

The Johns Hopkins University

I t is eminently clear that the delivery of health care in America is changing rapidly. Indeed, the transformation is not unlike the epochal change experienced by America's transportation system with the advent and spread of the steam engine. The academic health center (AHC), arguably the primary engine of growth and progress within the nation's health care system over the last century, is now at risk of being either supplanted by the new engine of managed care or relegated to the role of just another Pullman car on the managed care line.

But the health of the nation cannot afford to see the academic health centers supplanted or constrained to a very limited role. AHCs are the nation's best resource for medical education and training, scientific discovery, technological innovation, and advanced patient care. They are not simply centers of excellence. Academic health centers are what make centers of excellence possible.

If that is so, if AHCs are indispensable to the nation's health care, why do they find themselves under such scrutiny? And how might they be reengineered into strong new engines, coupled to managed care, and pulling the health care system effectively and powerfully into the twenty-first century?

The Challenge of Price Competition

The managed care "steam engine" is a response to the new, fiercely price-competitive health care marketplace. Past mechanisms for financing health care, combined with scientific advances and the demographics of an aging population, have created a cost spiral that endangers the viability of federal sources of funding for the nation's health care needs. Ironically, that same system has left many Americans without universal, affordable health care.

The development of a price-competitive marketplace for the management of people's health, often on a capitated basis, has emerged as a mechanism for containing costs while broadening health care coverage. Whatever its merits or limitations, the new marketplace presents special challenges for academic health centers.

Performing A Broad Social Mission. . .

Compared to virtually all other institutions in the health care system, academic health centers are charged with a broader social mission. They are not simply buildings with beds for the sick. They are the sources of the nation's supply of medical personnel; they are

the engines of discovery and innovation for the next generation of disease
prevention and cure; and, importantly, they are the largest providers of
uncompensated care for the nation's
poor, uninsured, and underinsured.

> **Academic health cen-
> ters are not simply cen-
> ters of excellence. They
> are what make centers
> of excellence possible.**

These missions extend deeply into the
health care capacity of the United States, and do not map well
onto a health system operated solely on price-competitive terms.
The wave of managed care has displaced the AHCs' paying patient
load and left behind its uncompensated patient base, without
thought to the consequent ability of the centers to continue as
financially viable institutions.

. . . With an Inherently High Cost Structure. . .

Given that broader mission, academic health centers have inherently higher cost structures than more "price-competitive" medical
institutions. For example, it is estimated that the full cost to the
AHCs in terms of lost clinical productivity associated with the
educational functions of faculty is nearly $2 billion per year across
the nation's 126 medical schools. Many additional billions in related costs are incurred by teaching hospitals in areas such as faculty
clinical research. Hence AHCs are not able to compete in cost-dri-
ven terms because they must build into their rates the costs of
medical education, uncompensated care, basic and clinical
research, and specialized care programs such as trauma, burn, and
advanced cancer centers.

. . . And Little Experience In Risk Management

The emphasis on price competition has led to the rapid development and consolidation of very large and capital-rich managed care

> Academic health centers are charged with broad social missions. These missions extend deeply into America's health care capacity and do not map well onto a fiercely cost-competitive system.

organizations in the health care system. To be a competitor in this arena, a health care facility or system must be able to manage the health care of hundreds of thousands of people. In turn, the institution must possess the ability to be both the insurer and the health delivery manager for these large populations. Risk management is the determinant of institutional viability, involving enormous stakes for the health care underwriter, provider, and manager.

Academic health centers, untutored in these approaches, find it difficult to adapt to these new roles. Despite being the centers of scientific innovation, AHCs are also often slow, inefficient, and sometimes stubbornly traditional. Their management and information systems are largely out of step with the streamlined systems of managed care organizations and networks. Hence their ability to meet new and higher risks with agile management of overall programs, information, and patient services is exceedingly limited.

Wanted: A Level Playing Field

Although academic health centers are clearly disadvantaged by the current price-competitive marketplace, they are not paralyzed. AHCs have the freedom to restructure management and operations, become more cost-sensitive, and maintain their commitment to public service missions of professional education, indigent and specialized care, and medical discovery and innovation. In order to succeed, however, AHCs require a level playing field. Some system for compensating AHCs for the impact of these missions on cost structure must be developed if the market-driven "competition" in health care financing is to be fairly played out across the nation.

In 1994, every proposal that passed a Congressional committee contained provisions for all-payer contributions for the academic health centers' educational mission. Such federal provisions would have uniformly leveled the playing field. Needless to say, the proposals died with the collapse of Congressional legislative efforts to fomulate a comprehensive health reform bill.

> A new reality is upon us and cannot be avoided. Financial support for academic health centers must be sought from sources other than the federal government.

Who Will Pay?

The burden may now rest on the states to take up these important issues and to try to develop goals and policies that can work for their citizens. Some states are moving forward. In Maryland, for example, efforts are under way to provide an affordable health insurance option for small employers and individuals who cannot otherwise afford such coverage. This group now makes up the majority of the uninsured and underinsured in Maryland. It is a similar scene in many other jurisdictions.

Even in Maryland, however, the degree to which government will assume the financial burden for leveling the playing field is still far from attracting adequate attention, let alone being resolved. Electorates remain woefully uninformed. Public policy is unformulated. For its part, managed care has not stepped forward to pick up a share of the financial burden for medical education and research.

Yet a new reality is upon us and cannot be avoided. Assuming that the educational, research, and indigent-care roles of AHCs are to be continued, to the extent that heightened competition reduces clinical cross-subsidies for these functions, financial support must be sought from other sources.

A Link to Managed Care

In the meantime, academic health centers around the country are bearing the burdens of service missions while striving to respond to the immediate financial and management imperatives of the new environment.

> **The real question is whether academic health centers can survive their own restructuring.**

The challenge for academic health centers is to convert into, affiliate with, or create within themselves managed care operations that can compete on a capitated basis for the care of large populations. This will require establishing close working relationships with communities, political leaders, HMOs, hospitals, and other community providers to maintain public service missions in the context of market participation.

The AHCs that succeed will be those that redesign faculty practice plans and hospital services into integrated group practices, networked ambulatory care facilities, and affiliated practice arrangements. In turn, the lines of responsibility and authority within the centers will have to be centralized if the resultant services are to remain price-competitive.

Implications for the Academic Function

Any successful organizational change will have profound implications for the faculties of academic health centers. The core of academic faculty, particularly those whose work is dedicated to research, discovery and teaching, likely will have to be smaller. New job descriptions will emphasize clinician functions.

Independent patient billing will be a thing of the past. Service efficiency will rule, together with accounting and accountability. Indeed, physicians may find themselves accounting for their work hours and days in ways now familiar to lawyers. The teaching function will need to be reorganized and segregated. New types of skills will be needed. Leaders of clinical departments will be people with superior management skills and experience in managed care delivery systems. New skills at the medical level will also be required, with greater emphasis on primary care and ambulatory settings.

The Question: Will AHCs Survive the Process of Change?

It remains to be seen not just how, but whether the values of the academic environment can be preserved within a price-competitive marketplace. These values have provided fertile ground for the spectacular advances of this century in the biomedical sciences and in many other disciplines. There is reason for optimism because AHCs have faced and conquered huge challenges in the past—the new biology of molecular and structural biology; the nature of the human body and the application of that knowledge for human good; and imaging, optics, robotics, nanotechnology, and biomaterials, to name but a few.

Academic health centers now face their greatest challenge: to survive their own restructuring. The differences in structure and in function between market-driven managed care and educational organizations are real and significant. These differences must be faced, aligned, and linked, in the interests not simply of the centers themselves, but in the interests of the nation's health care services for generations to come. ■

3

Societal Dilemmas

Humankind does not lack for challenges. Unfortunately, the future of global health is similarly blessed. Old diseases long ago thought confined to deep channels of history, rise up again and surge over weakened public health levees. New diseases emerge from changing interactions between humans and the environment. A growing global population strains health resources and, combined with greater global mobility, provides a continuous pathway for the global spread of disease. And, perhaps the most heinous of all, disease itself is harnessed by humans as a weapon of mass destruction.

In all of these dilemmas, the challenge is global. The time is long past, if indeed it ever existed, when one nation or one people could isolate itself from the world, address only its own public health concerns, respond only to its own internal disease priorities. The essays that follow pivot on this common theme: there is a unity to the dilemmas that confront global health, and there must be an equal global unity of purpose—among nations, across disciplines and between science and public policy—in responding to the complex challenges that these dilemmas pose.

Emerging and Resurging Infectious Diseases: The New York City Perspective

MARGARET HAMBURG

Commissioner of Health for New York City

Based on remarks delivered at the Science Policy Association of the New York Academy of Sciences on January 26, 1996. In October, 1997, Dr. Hamburg became Assistant Secretary for Planning and Evaluation at the Department of Health and Human Services.

Few public health issues are as timely or as important as that of emerging and resurging infectious diseases. Certainly, few have such massive scientific and societal implications, both at the local and at the global levels. Emerging and resurging infectious diseases are not just the stuff of science fiction. They are not hypothetical threats. They are a present, real, and very imminent danger.

Mortality from infectious diseases has been increasing in the United States in recent years. In a large, dense urban area like New York City, it is possible for a previously unrecognized organism or unexpected disease to insinuate itself into many aspects of societal life. New York City, for example, is the epicenter of the AIDS epi-

> Emerging and
> resurging infec-
> tious diseases are
> not just the stuff
> of science fiction,
> they are a present,
> real, and very
> imminent danger.

demic in this country. To date, the city has recorded more than 80,000 cases of AIDS. Approximately 130 deaths each week are attributable to this disease, and it is now the leading cause of death in the city for both men and women between the ages of 25 and 44.

In fact, we are seeing a dramatic increase in infectious disease threats, including newly recognized diseases like hantavirus pulmonary syndrome, *E. coli* 0147:H7, cryptosporidiosis, and erlichiosis, as well as the resurgence of old diseases like tuberculosis and rabies.

The Case of Tuberculosis

The cycle of tuberculosis that New York has recently confronted presents an example of how a resurging disease, made more complicated by multiple drug resistance, can quickly take hold. After decades of decline, the incidence of tuberculosis increased 132% in New York City between 1980 and 1990. We did not lack medical knowledge about tuberculosis. We did not lack the tools to prevent and control the disease. Indeed, we had demonstrated historically that significant strides could be made against tuberculosis. What happened?

Three trends came together in the city, with just the right mix to feed the resurgence. On the one hand, New York experienced increasing numbers of residents in poverty, in crowded living conditions, or in incarcerated circumstances. Many of those residents were from nations where tuberculosis is a prominent disease. Second, the co-epidemic of AIDS made many individuals more vulnerable to tuberculosis. Third, the public health infrastructure that, decades ago, had been critical to controlling the disease had begun to be dismantled. In effect, we had dropped our public health guard.

Over the past few years, New York has once again turned the tide of tuberculosis. In the last two reported years, case rates have declined by 22%. There have been dramatic declines in the rates of multiple drug resistant tuberculosis as well. Further decreases are likely. What lessons have been learned? The key to reestablishing control did not lie in complex, highly technological approaches to the problem. Rather, basic public health principles and strategies once again proved to be the strongest medicine. Aggressive case finding, appropriate and swift diagnosis and treatment, outreach with the use of directly observed therapy—all ensured that persons who needed treatment were identified, received medication, and completed their therapy. This strategy, combined with educating the public and, importantly, health-care providers to be aware of and vigilant toward the tuberculosis resurgence, has proved effective. Nevertheless, there is danger in success just as there was crisis in failure. If current success breeds complacency, the cycle will begin anew.

> After decades of decline, the incidence of tuberculosis in New York City increased 132% from 1980 to 1990, despite the fact that we had the medical knowledge and technological tools to prevent and control it.

Moreover, problems revealed in the reemergence of one disease may be reflected in other diseases. Although drug resistance became a theme in media coverage of the tuberculosis problem, it is hardly limited to that disease. Public health authorities worldwide are seeing a resurgence in drug-resistant organisms; unfortunately, New York City is in the forefront. For example, in 1991 a survey found that vancomycin-resistant enterococcus was fairly prevalent in the city; in the last year, every hospital has reported the resistant strain. Similarly, penicillin-resistant pneumococcal dis-

> Despite the greatest concentration of sophisticated biomedical research and clinical institutions found anywhere in the world, New York City is still vulnerable to infectious disease.

ease is increasing, leading to a need for more in-patient care and making treatment more expensive.

The Global Link

With rapid and widespread global transportation, the problems New York faces cannot be limited to the nature of its internal systems or populations. New York is an international city, and its public health problems will reflect that fact.

The outbreak of Ebola virus in Africa and the reports of pneumonic plague in India were taken as serious threats by the public health authorities of New York City. Working with the federal government and the Centers for Disease Control, the major airports were monitored, and potential cases screened. The medical community was alerted and educated about symptoms. Potential respiratory isolation rooms in hospitals throughout the city and triage units were identified. Fortunately, no cases were detected. But, a great deal was gained from the process of emergency preparedness; in the future, the process will benefit from the past drill.

Why Is New York Vulnerable?

Despite such preparedness and despite having the greatest concentration of sophisticated biomedical research and clinical institutions in the world, New York is still vulnerable to infectious diseases. Many of the city's characteristics combine to raise its risk profile for the emergence of diseases of infectious origin. The pri-

> We cannot recognize the unusual if we do not engage in routine surveillance.

mary requirement for the occurrence of a new epidemic is not the appearance of a new microorganism but the occurrence of conditions that allow for the rapid dissemination of the disease. Unfortunately, virtually all of those conditions exist in New York City.

The city has an extremely high population density. In many communities that density is compounded by severe poverty. Unlike many municipalities, New York has an unfiltered water supply. Moreover, it is a major port of entry into the country for international travel and commerce and has a large and growing immigrant and refugee population. New York is also host to a large number of transients ranging from tourists to visiting students, scholars, and business people, as well as domestic and blue collar workers who come from across the nation and around the world. Greater numbers of people are placed in closer contact with vectors and pathogens. For New York, the line between domestic and international health has been virtually obliterated.

What Can Be Done?

The primary weapons at our disposal in this escalating war on emerging infectious diseases are the core functions of public health. Health status monitoring and disease surveillance, investigation and control of disease, protection from environmental hazards, health education and disease prevention, outreach, public and professional mobilization, and direct clinical services for disease control—all are crucial elements. But, the core is surveillance. We cannot recognize the unusual if we do not engage in routine surveillance. Surveillance provides the data about the common that makes the uncommon stand out. It also provides the means for evaluating the effectiveness of interventions.

All of these public health functions are becoming both more

important and more endangered. Even as the threat grows, the first line of defense weakens. In the many debates over health care and health-care reform of the last few years, public health has been resoundingly missing. With the nearly one trillion dollars spent in the health-care system of the United States, less than one percent is allocated to performing vital public health functions, and that percentage has been declining in the last decade. Without addressing this core capacity of public health in the nation, the prospects on the infectious diseases battlefield are poor. ■

Infectious Diseases and Nature's Revenge

Based on remarks delivered at the Science Policy Association of the New York Academy of Sciences on January 26, 1996.

JOSHUA LEDERBERG

University Professor
The Rockefeller University

T here is an important philosophical perspective that must be appreciated to understand the current re-emergence of infectious diseases and to begin to think about a response. One could entitle it "Nature's Revenge" or "Nature versus Art."

We are, essentially, an unnatural species. With the evolution of intelligence, the transmission of culture, the development of technology, and the building of social institutions, humans are unique among all creatures on earth. We have expanded far beyond the bounds provided by nature. We use technology to augment our numbers, approximately a thousand-fold over what a natural species of our capabilities would have achieved.

We have created artifacts—crowded cities, health-care systems, medical technologies—to support our lives. We also have an extraordinary propensity for travel. A million passengers a day cross interna-

> The human species has used its technology to augment its numbers approximately a thousand-fold over what a natural species of our capabilities would have achieved. We are victims of our own success.

tional boundaries by air alone. Armies, refugees, truckers, and the normal pedestrian all contribute to an endless mixing of the species. This makes a wonderful breeding ground for nature's germs.

In the past, we have used our technologies and artifacts to triumph over these germs, over infectious disease. As we have become complacent, however, what we call the emergence of infectious diseases is really the prospect of regression, of falling back to the time before technology gave us our first victories. We have nodded off on our watch. But, the bugs aren't sleeping. A very powerful process of natural selection is operating, and they will exploit every available niche that we provide. And, without applying technology and defenses that we know over time to be successful, we will provide more and more niches as time goes on.

Having once committed ourselves as a species to a highly unnatural course of development, one that meant a life span of thirty or forty years beyond the years of procreation, having developed all the artifacts of civilization, having decided we are going to rely on our wits for survival, we cannot

> We may drop our vigilance against infectious diseases, but the bugs are not sleeping. A very powerful process of natural selection is operating, and they will exploit every imaginable niche that we provide.

afford to drop our vigilance. No other creature on earth can compete with our species, except ourselves and microbial predators. We are in a necessarily constant struggle with those predators, and we had best not drop our guard. ■

Facing the Priority of Disease Surveillance and Control

Based on remarks delivered at the Science Policy Association of the New York Academy of Sciences on January 26, 1996.

STEPHEN S. MORSE

Project Manager

Defense Advanced Research Projects Agency

 n the last few years, humanity has had grim reminders about how fine is the line between progress and the unknown.

Predators in the Global Village

The May, 1995 telephone call will stay with me for some time to come. On the other end was a reporter who wanted to know if I had heard about a group of nuns who had returned to Italy from Zaire with stories of the deaths of several colleagues who had succumbed to a devastating infection. The ultimate medical and epi-

> **There is no longer anything exotic about "exotic" diseases.**

demiological verdict was Ebola, and the world gave a collective shudder. Although the returning nuns were not infected and there was likely never a real threat to the United States, the episode, as it was followed in excruciating detail in the press, served as a signal of how much we have become a global village.

Mere months passed, and attention turned to the outbreak of plague in India. With its reminders of the Black Death of the fourteenth century, the outbreak sent another collective shudder around the globe. Other stories of the resurgence of cholera, tuberculosis, and the like also regularly make headlines.

Of course AIDS, a disease that was totally unknown at the end of the last decade, has now claimed countless productive lives and cost billions of dollars in direct outlays for medical care and in lost productivity. In Thailand, an important trading partner of the United States and a force for regional stability in Southeast Asia, there are villages in the northern regions with zero negativity rates for H.I.V., the virus that causes AIDS, with full-blown AIDS approaching 20 to 25% in men.

Exotic No Longer

These outbreaks, and their impact on humanity, by now represent a familiar litany. There is no longer anything exotic about exotic diseases. The world is a small place, and getting smaller and more interconnected over time. Of course, there is much that is salutary about this development. It leads to opportunities for people to interact and get to know and understand one another, for trade to expand, for technology to spread. It is also an opportunity for microbes, wherever they first appear and wherever they may now be living, to travel rapidly with people and goods around the world. The World Travel Organization has announced that in the first quarter of 1995 international travel increased 6% over the year before. And that was just for the first quarter! The millions of daily international trips—by boat, by air, by rail, by foot—are increasing every day and will continue to increase.

What is happening here? Why do we face resurgent diseases and entirely new infectious agents? If we look closely at the sources and spread of these diseases, we can see a two-step process.

Factors Promoting Resurgence

The first step is the introduction of a new agent, an agent that is unknown to us but that is not new in nature. In most cases, this is an agent that already exists in the environment as part of the vast biodiversity of microbes throughout the world. Often the most novel infections, such as plague, are zoonotic. They are infections of other species introduced into the human population, usually as a result of changing ecological or environmental conditions that have given some population, somewhere in the world, the opportunity to come into contact with the species naturally carrying the infection. This process represents a very strong argument for maintaining a high level of environmental sensitivity. Major economic, infrastructure, or development projects that change the environment often have public health effects as unanticipated consequences. Indeed, "unanticipated" may no longer be the correct term. There is plenty of reason to anticipate the consequences. We now have numerous examples of dam building, forest clearing, and agricultural expansion that result in the introduction of previously unrecognized infections into the human population.

Having been introduced, then, the infection requires a second step. It must become established and disseminated within the human population. Here too, the conditions of modern life pro-

> **The declining state of public health infrastructure will offer opportunities for infections to gain a foothold before they are even recognized by the health system.**

mote that process. The case of H.I.V. provides a tragic example. Probably first introduced into an isolated human population from some other species, H.I.V. propelled itself to infamous stardom by the many factors that allowed it to spread. Movement of peoples from isolated areas to rural areas and then to urban areas, largely because of economic considerations, promotes dissemination. International travel then speeds the process.

Public Health as Firewall

It is not just the changing economic base of many countries nor increases in travel that contribute to the problem. Declining investments in public health infrastructure remove the firewall from the dissemination process. Without public health capacity and active surveillance, the infection can gain a silent—and often lengthy—foothold in a population before it is discovered. At that point, control becomes infinitely more difficult.

If there is one thing that every group dealing with the many aspects of disease control agrees on, it is the imperative of disease surveillance, at home and abroad. The United States, with many of the world's more significant public health resources, can and must play a leadership role.

"Ah, yes," it is often said, "but this costs money. Resources are limited. How much would such careful attention to surveillance cost?" The answer is, relatively little. Consider the irony that Dustin Hoffman was paid more for making the movie *Outbreak,* Hollywood's cinematic spin on the emerging infectious diseases problem, than the annual budget of the entire special pathogens branch of the Centers for Disease Control. He could have funded a year's worth of global surveillance out of his salary!

> **Dustin Hoffman was paid more for his role in the movie *Outbreak* than the entire budget of the special pathogens branch of the Centers for Disease Control. He could have funded a year's worth of surveillance out of his paycheck.**

The scientific resources that we can bring to bear on the disease control problem are greater now than ever in history. Through telecommunications technology and the Internet, the possibilities of cheap and effective global communication—a keystone of surveillance—are truly pathbreaking. Even now, we have developed an experiment on the Internet in which an electronic conference regularly carries news about infectious disease surveillance to over 3,500 subscribers in 125 countries at a very low cost and without the need for massive infrastructure.

War It Is

Scientific knowledge and agreement on priorities and methods is not the problem, however. It seldom has been. What is lacking is political will. The political process must establish the means for decisive action to be taken. There is every reason to act. We are fighting for our survival against implacable enemies that have been around since the dawn of existence. Indeed, some of them are the earliest forms of life. They have learned every evolutionary trick in the books. They endure and adapt through endless cycles of history.

The human fight against these enemies is neither obscure nor subtle. They are the predators; we are their prey. In maintaining constant vigilance and in fighting with the totality of our scientific arsenal, we are not engaged in the moral equivalent of war. This is the real thing. ■

Organizing to Address Global Issues

Based on remarks delivered at the Science Policy Association of the New York Academy of Sciences on January 26, 1996. In January, 1998, Mr. Wirth became president of the U.N. Foundation in Washington, D.C.

TIMOTHY E. WIRTH

Under Secretary of the Department of State

T he problem of confronting the resurgence of infectious diseases is part of a new set of global issues for our national security. How have we dealt with national security issues in the past?

Something Old, Something New: The Global Approach

Historically, we have organized ourselves on three dimensions to address national security issues: political, military (especially related to arms control), and economic. The Clinton administration has introduced a fourth approach to this traditional mix, an integrated approach to global issues. Within each mission agency—the Department of Defense, the C.I.A., the State Department, the National Security Council, the Agency for International Development—new clusters of capability have been formed to address global issues and to begin to focus the attention of traditional bureaucracies on these issues.

> **Emerging and resurging infectious diseases are at the top of the list of national security interests.**

In the State Department, such global issues fall into four broad categories: (1) democracy, human rights, and labor; (2) refugees, population, and migration; (3) narcotics, terrorism, and crime; and (4) oceans, environment, and science. Giving this global approach visibility and accountability within the larger organization emphasizes that these issues are important and need priority attention.

Setting Priorities

Of course, the problem of setting priorities remains. Clearly, innumerable problems could be addressed in each of these categories. Which of these are at the top of the policy list?

From the perspective of the Clinton administration, there are five key global priorities. First is climate change. This is a major area of concern and attention across the global groups in the various government agencies. Second is narcotics, a dangerous plague for this country and many others. Third is population, obviously a continuing area of concern. Fourth is food security and the relationship of food security to long-term political stability around the world. Fifth is biological diversity and infectious diseases.

> **It is not always easy to move the machinery of government around to address new concerns from new perspectives.**

A clear sense of urgency exists in this administration on all of these priority issues. It is not always easy to move the machinery of government around to address new concerns from new perspectives. But, gradually, progress is being made.

The Importance of Partnership

> Infectious diseases and public health are truly nonpartisan issues. Microbes do not respect party lines.

Obviously, the United States cannot resolve these global issues alone. We are building active partnerships with other world leaders to ensure that a concerted effort is being made. For example, we have a new common agenda with Japan to work together on all five of the priority clusters. In response, the Japanese have increased their commitment to global programs for population and AIDS from $40 million to $400 million per year.

The development of the Committee on International Science, Engineering, and Technology (CISET) report on global infectious diseases in September of 1995 is an example of a new spirit of collaboration within the U.S. government. The analysis and policy recommendations reflect broad and deep collaboration among a wide range of U.S. agencies to attempt to develop common priorities and directions. Similar reports have been completed on AIDS and on food security.

Microbes Meet Politics

It is very important to emphasize that these issues, including priorities in infectious diseases, are truly nonpartisan issues. Microbes do not respect party lines. We have an immediate opportunity to educate the political leaders on both sides of the congressional aisle about the importance of the disease control imperative. This global priority is critical for the United States, and science can take the lead in ensuring that this is understood and acted upon. ∎

Reproductive Health & Reproductive Products: Politics, Products, and Processes

Based on remarks delivered at the Science Policy Association of the New York Academy of Sciences on April 15, 1997.

MARGARET CATLEY-CARLSON

President, Population Council

The Perfect Contraceptive

In a simplistic vision that never had validity, there always existed a wistful hope that soaring demographic curves in developing countries could be intersected by the introduction of the perfect contraceptive. We now know that all contraceptives are, in some way and in some markets, niche contraceptives. The Cairo Conference has added both convergence and complexity to the way the world now looks at the role of contraceptive services in a broad menu of population assistance programs. The new paradigm carries the name "Reproductive Health."

> **The population community joined with women's health advocates in affirming that a client-centered approach, and a focus on the situation of women, was to become the *global* organizing principal for reproductive health.**

Like all new developments, it is not totally new—especially in most parts of the industrialized world where contraceptive services and reproductive health care decisions have traditionally been largely an issue between women and their healthcare provider. The element of novelty here—and the main reason the Cairo Conference represented a step forward—was that the population community joined with women's health advocates in affirming that a client-centered approach, and a focus on the situation of women, was to become the *global* organizing principle for services in the years ahead. But the message from Cairo applies to all women, not only those in poor countries. All women need choices, medical abortion, STD protection, well-trained providers, and access to good and correct information.

As part of this impetus, family planning organizations and many practitioners around the world are now struggling to work through how contraceptive services must be reconsidered in order to support and improve women's health. This should include, above all, the client-centered provision of services with necessary cross-referencing of the client's life situation, risks, etc. With respect to technology development, intensified attention must be given to the degree to which given contraceptives—either through their intrinsic properties or by the mode of service provision—will enhance or detract from client reproductive health. Specific information about the woman and her life situation is key to determining this.

These implications apply to practitioners in developed and developing countries alike. Those working with poor women or ethnic minorities, those trying to integrate the reality of violence against women (including Female Genital Mutilation) into clinical practice are all part of the movement toward a new integrated social medicine. Promoting reproductive health in the 1990s

> **Contraceptive technology is not neutral in application.**

requires an ever-increasing sensitivity to the actual conditions of sexual relations and the power balance between the sexes. Contraceptive technology is not neutral in application—its effectiveness, health impact, and safety depend directly on the service context and the quality of intimate relationships.

For more than four decades, the Population Council, an international non-governmental organization, has performed research in the population and reproductive health fields. Six Council products are available to women around the world for contraceptive and reproductive care; these are confirmed by a wide global experience in the diverse cultural and social contexts.

The Perfect Development Process

In 1979, Carl Djerassi wrote, "Except for a few special vaccines, essentially all modern prescription drugs have been developed by pharmaceutical companies. I know of no case in which all the work (chemical, biological, toxicological, formulative, analytical and clinical studies through Phase III) leading to government approval of a drug was ever performed by a government laboratory, a medical school, or a not-for-profit institution." (*Politics of Contraception,* 1979.)

But the Population Council does all of this—all of the above. We do not market and we do not manufacture but we do everything else.

The Council entered into the contraceptive development busi-

> **The objectives of the Population Council's contraceptive program are to increase the number of contraceptives in family planning programs, to complete the development of new contraceptives, and to identify new leads.**

ness when large pharmaceutical companies backed out of the field after the development of the pill. Common wisdom has it that there once were 13 sizable pharmaceutical companies working in the contraceptive/reproductive health area. Today there are five. Why did they down tools? The belief was that it would be difficult to improve on the pills followed by long lead time relative to patent protection, declining public tolerance for the risk, and difficulty in obtaining product approval. For a brief while, when sales of a truly new drug delivery system (Norplant®) were zooming and media attention was high and positive, there were some signs of a resurgence in interest. The story of Norplant® to date has not been such as to sustain and nourish such an interest.

The Institute of Medicine has theorized that there is no single factor to explain why companies have retired from the field— "Liability, economic factors, the size and character of an industrial firm, regulations, politics and ideology, attitudes and behaviors of individuals" all join to reinforce the trend away from investment in this area.

The WHO Special Program of Research and Development in Human Reproduction began in 1972; WHO with Conrad have grant-making programs to foster the science that supports contraceptive development. Family Health International (which dates back to 1971) and Path work in this area, too. The Council probably accounts for about 20% of global research in this area, although true figures including pharmaceutical company research are notoriously difficult to obtain.

The broad objectives of the Council's contraceptive development program are: to increase the number of contraceptives that are registered for use in family planning programs; to complete the development of new contraceptives with specific advantages over existing contraceptives; and to identify new leads. The Council gives greatest emphasis to those contraceptives suited for distribution to family planning programs in developing countries.

> In interesting private industry in contraceptive markets, indemnification is a major issue as are the mounting of substantial resources, appropriately sized marketing, and distribution efforts.

In-house activities provide for laboratory studies of animal physiology and toxicology as well as the formulation and manufacture of devices to be tested. Clinical trials are initiated subsequent to approval by regulatory authorities. The principal instrument for the Council's clinical trials in the International Center for Contraceptive Research (ICCR), which was established in 1970. ICCR consists of investigators from several countries around the globe who evaluate different methods.

If no untoward effects are identified in small-scale clinical trials or additional toxicological trials, clinical trials are expanded to establish the dose of drug required in humans. This is followed by the development of a large scale manufacturing process and Phase III clinical trials. Then, if safety and efficacy are proven, the submission of an New Druge Application (NDA) with the U.S. Food and Drug Administration (FDA) is made—a necessary step to permit the use of these products in USAID family planning programs.

The process of patenting and licensing is not an easy process. Experience with innovative systems such as implants and rings shows that there is a better prospect of success if a manufacturer has been secured. This means that we must approach a potential marketer with a whole package: source of active ingredient, manufacturer for the drug delivery system, regulatory package of toxicology, pharmacology and clinical trials to support safety and efficacy.

For the same reasons that caused major pharmaceutical companies to move out of research and marketing in this area, many of our discussions have been with very small pharmaceutical companies. The experience is mixed: indemnification is a major issue as are the mounting of substantial resources, appropriately sized marketing and distribution efforts.

New Products and a More Perfect Future

Expanded contraceptive choice centers on the contraceptive need of women. Expanded choice means that technology must be developed to meet the contraceptive needs of currently unserved or underserved groups. These needs encompass methods that are under the user's control: methods for men; methods for nursing women and teenagers; and methods that can be used without altering women's menstrual cycles.

The leitmotif has moved on from contraception alone to more choice: products that assist reproductive health at all ages, products that help with disease prevention, and safer motherhood, including medical abortifacients, are now needed. The Council is working on all of these. Women's needs and desires about contraception have changed over the past 20 years. IUDs and Norplant® were developed in specific response to the ideal of a long-acting method that was reversible but required little attention from the user. Now the needs are different.

The contraceptive ring is a doughnut-shaped drug delivery system inserted into the vagina. The ring contains a steroid or steroids—a progestin or a progestin plus estrogen—which is slowly released into the vagina and then into the bloodstream. The method's attractiveness for providers and users lies in the fact that women can insert and remove the silicone rubber ring themselves. Another positive aspect is that in-depth studies among ring users and women not using these devices show that ring use does not increase the incidence of vaginal irritation. Several variations of rings are being developed:

• Progestin-only ring which, because it contains no estrogen, can be used by lactating (nursing) women as well as by women who have weaned their children. Therefore, a woman does not have to switch methods when she stops nursing and starts ovulating.

• Combination rings which bring the benefits of different hormones simultaneously. The postmenopausal rings of estradiol with or without progesterone as hormone replacement therapy in postmenopausal women. The use of estrogen is expected to alleviate symptoms associated with menopause, and to decrease the risk of osteoporosis and cardiovascular disease.

• The levonorgestrel delivery system, available as Mirena® in 14 European countries, developed by the Population Council and the Finnish pharmaceutical company Leiras Oy, involves the addition of a sustained delivery of levonorgestrel to a T-shaped intrauterine device for local effect on the endometrium. Mirena® reduces blood loss and protects against anemia. The Council is continuing studies on health benefits including fibroma and post abortion use.

Implantable Contraceptives for Women
The Norplant® implant is a five-year, highly effective female contraceptive consisting of six, small, flexible capsules that are inserted under the skin of the inner side of the upper arm. The capsules continuously release a low dose of synthetic progestin into the bloodstream to prevent pregnancy. The Norplant® implant has been

> **Council scientists are continuing to screen agents that have potential microbicidal actions to prevent sexually transmitted diseases, including AIDS.**

approved in dozens of countries and is available in many countries including the United States, the United Kingdom, and the Scandinavian countries.

The levonorgestrel 2 rod implant (aka Jadelle) is a highly effective female contraceptive consisting of two small, flexible rods that are inserted under the skin of the inner side of the upper arm. This is a more efficient implant system, and is expected to be easier for healthcare providers to administer. The implants release continuously a low dose of the progestin levonorgestrel into the bloodstream to prevent pregnancy. To date, the levonorgestrel 2 rod system has shown to be extremely effective for a three-year period. Preparation of an NDA for Norplant® II as a three-year method is complete and the Council has received NDA regulatory approval from the U.S. Food and Drug Administration last year. However, the Council will continue studies through 1997 to determine the 2 rod system efficacy beyond three years.

Protection Against Disease Transmission: Microbicides

Council scientists are continuing to screen agents that have potential microbicidal actions (with or without spermicides) to prevent sexually transmitted diseases, including AIDS. Laboratory teams are testing sulphated polysaccharide for anti-HIV activity with new in vitro and animal assays. Many compounds in this class are available commercially and are generally inexpensive, safe, and watersoluble-desirable characteristics for a vaginal formulation. In addition to sulphated polymers, Council scientists are developing other leads for use in vaginal formulations.

To measure women's acceptability of this kind of product, a

study in five clinics in both developed and developing countries is underway to determine which traits of currently available vaginal spermicidal products positively or negatively influence their use. We are also working with women's health activist groups to develop common ground on this project.

Safer Abortion

In May 1994, Roussel Uclaf donated the US patent rights for mifepristone, formerly called RU 486, to the Population Council. In autumn 1995 the Council concluded a Phase 3 clinical trial for mifepristone as an abortifacient with approximately 2100 women in 17 clinics across the United States. The trial has tested mifepristone (used in conjunction with an oral prostaglandin, misoprostol) in 2,100 women with amenorrhea of less than 64 days. The Council has received an approvable letter from its NDA to the Food and Drug Administration, selected a manufacturer and distributor for the drug in the United States. Anybody watching this saga will know it has been far from easy.

But we continue on . . . now, how to get more entities interested? That is the task ahead. ■

Endnote
1. Polly F. Harrison & Alan Rosenfield, *Contraceptive Research and Development: Looking to the Future* (Washington, D.C.: Institute of Medicine, National Academy Press, 1996)

Contraceptive Research: The Once and Future Revolution

Based on remarks delivered at the Science Policy Association of the New York Academy of Sciences on January 26, 1996.

ALLAN ROSENFIELD

Dean, School of Public Health
Columbia University

I n the early 1960s, the contraceptive pill was introduced and the IUD reintroduced to the marketplace, launching what some have called the "contraceptive revolution." Millions of women around the world, for the first time, were given the means of preventing pregnancy that was separated from the act of intercourse. That was nearly four decades ago.

A Second Revolution?

But the revolution is not over yet. Millions of individuals still do not have access to contraceptive technology. This a huge, urgent, and unmet need. Yet it is not clear that society is prepared to respond. Societal, political, and financial factors have stalled research and development efforts. Most new contraceptives and

> There is a huge, urgent, and unmet need for contraceptive technology. But it is not clear that society is prepared to respond.

most development activity is focused on modifying existing products and approaches. The hormonal concept underlying contraception was introduced with the pill in 1960. We have improved the pill. Hormones, in a long-acting formulation, now can be delivered by injection. We have developed an entirely new delivery system, the implant. But at their core, all of these variations are based on a hormonal approach. Current efforts are not creating anything fundamentally new.

In 1996, the Institute of Medicine (IOM) was asked by the Rockefeller Foundation, the Mellon Foundation, the National Institute of Child Health and Human Development (NICHD), and the U. S. Agency for International Development (USAID) to gather a group of experts, from multiple disciplines, working in universities, government, the pharmaceutical industry (including small biotechnology organizations) and consumer advocacy groups, to explore the scientific and practical potential for a second contraceptive revolution, a search for new and innovative approaches to contraception.

As is the practice with IOM committees, some members worked directly in this field and others worked in related areas. Before the committee was willing to move forward there was a detailed review of the evidence as to whether or not there really is a need for new contraceptive technologies. If such a need could be identified, then the questions were what is the state of the science, and what are the prospects for rekindling industrial interest and investment in contraceptive research and development?

> Between 120 million and 230 million women in the world are at risk of an unintended pregnancy at this very moment.

Is There a Need for New Contraceptive Technology?

Depending on the methodology used, between 120 million and 230 million women in the world currently are at risk of an unintended pregnancy. Between 24% and 64% of pregnancies worldwide are unintended at the time of conception. In the United States,

> Scientific advance has opened the way for a second revolution in contraceptive technology. None of the breakthroughs will be easy. And all will be expensive.

57% of the nation's 3.1 million pregnancies are unintended at the time of conception, and 45% of these (or some 1.3 million) end in abortion each year. There are over 50 million abortions worldwide annually, with an estimated 20 million of those performed illegally. It is estimated that between 70,000 and 120,000 deaths occur each year secondary to botched abortion procedures.

In addition, there are well over 300 million new cases annually of curable sexually transmitted diseases (STDs), including gonorrhea, syphilis, and chlamydia Fourteen million of these cases are in the United States alone. Further, of course, the tragic HIV/AIDS epidemic afflicts millions of people throughout the world.

These data clearly point to a need for more choices for contraceptive methods, as well as methods specifically related to protection against STDs. The failure of current methods may relate to misuse or non-use of available contraceptive technology due to side effects, to myths, or to misinformation from friends or the media. It may also result from lack of access to reproductive health services or to the fact that no contraceptive device, including sterilization, provides 100% protection. However complex the roots of the problem, it is obvious that an expanded variety of, and improvement in, contraceptive technology would be essential to substantially change the portrait painted above. In this search, the need to give attention to methods that are under the direct control of women was stressed, particularly in relation to STD protection.

Can Science Respond?

If the need is real, what is the state of science? Can recent advances in molecular and cellular biology produce truly new and fresh approaches to contraception and the protection against STDs?

In short, the news is good. In the short term, new and improved methods building on current technologies are possible. In the long term, new scientific approaches are opening the way to radically new contraceptive methods for men and women. It will be possible to focus at specific molecular levels on the processes of ovulation in women and of sperm maturation in men. Such specificity would allow for the development of approaches that have few systemic side effects. The potential for immuno-contraception, or a contraception vaccine, is also very real, albeit controversial.

None of these breakthroughs will be easy. All will be expensive. And all will take a number of years before they can be brought to the market. But scientific advance has opened the way for a second revolution in contraceptive technology.

The Critical Role of Industry

The question, of course, is who will lead? Translating science into cutting-edge products requires huge investments. Capital of the size required is only present in private industry. But the response from industry will contin-

ue to be tempered by the difficulties of translating need and scientific promise into a profitable and
safe market. The high costs and
risks of committing resources to
the development of any medical
technology are such that no firm
will undertake the commercialization of a product without the existence of a market of new con-
sumers able and willing to pay for the product. And there is little
incentive for the industry if a new method is a less expensive sub-
stitute for oral contraceptives, which have been a major profit area
for many companies.

> **With RU46, Roussell Uclaf
> decided it could not win
> whatever it did. The compa-
> ny sold the rights to the
> Population Council and
> closed the research pro-
> gram that had led to the
> development of U46 in the
> first place.**

For contraceptive products, return on investment is only the
first hurdle. Innovations in reproductive technologies face problems of liability, the regulatory process, politics and ideology, and
individual behavior. All are sensitive issues, and many are unique
to particular nations. The "global" market in contraceptive technology is very local indeed.

Two Chilling Events

These factors in general create a cautious environment for industry
exploring product potential in contraceptive technologies. But two
events, which took place during the deliberations that led to the
Institute of Medicine report, demonstrate the severity of some of
these issues.

Controversy over RU46 (mifepristone), the medical abortifacient developed and marketed by the French firm Roussel Uclaf
had a remarkable effect on the company. Those opposed to abortion were very aggressive in their opposition to this new drug,

> **Contraceptives perform a social function similar to that of vaccines, produce a similar social good, and should be afforded similar protection against liability.**

threatening, among other things, a nation-wide boycott of all of the company's products if RU46 was brought to the U.S. market. In reaction to the company's response to the anti-abortion groups, pro-choice groups became very active, including one group that arrived at the company's headquarters with petitions signed by some 100,000 Americans demanding that the company market the drug in the United States.

The result? The company continued its very limited marketing to France and a couple of other European countries and released the rights to the drug to the Population Council. It then closed the research program that had led to the development of RU46, and eventually shut down the company's entire endocrine research division. The company measured the intensity of the debate and decided that it could not win whichever decision it made. The industry does not like controversy, no matter their own particular views on such an issue. This episode has put a chill on the development of any contraceptive method that might also have the potential to be an abortifacient.

At about the same time another contraceptive technology controversy erupted, in this case an attack on Wyeth's contraceptive implant, Norplant, with complaints that the method was unsafe, relating to problems of removal, alleged immune and cardiovascular complications. Shortly after these issues received major media attention, sales of Norplant plummeted. A subcommittee of the IOM Committee on Contraceptive Research met to review the evidence about this method and concluded that Norplant was safe and effective. While removal of the Norplant rods had been difficult in some cases, where there had been adequate training, this complication was very rare. And a newer version, a two-rod method instead of the current six could be made available, but is not being promoted by the company.

Industry, participating in and/or observing these often impassioned controversies, is understandably reluctant to commit to the levels of capital investment that would be needed to fuel a second contraceptive technology revolution.

The IOM Committee Recommendations

In concluding its assessment, the Institute of Medicine report made a series of recommendations targeted at overcoming some of the barriers that impede industry from taking advantage of scientific breakthroughs in the context of clear market need.

First, a global contraceptive commodity program should be created. This central agency would provide contraceptives at a low cost to programs throughout the world, and hence would represent an immense procurement point, in turn making the market attractive to industry.

Second, in the United States greater effort should be made to encourage managed care to extend its coverage not only to surgical contraceptive procedures but to preventive technologies as well. Again this would increase the market through strengthening the flow of service financing.

Third, aggressive action should be taken to mitigate industry's liability concerns. The report recommended a Federal Protection Standard. If a company brought a product through the approval process of the Food and Drug Administration (FDA), and if FDA approved the product for sale, then the FDA would grant limited protection from liability to the company. This is similar to the protection given vaccines now, which are seen as a critical social product. The nation's need for vaccine production, the increasing risks of major liability for vaccine producers, and the consequent unwillingness of industry to stay in the vaccine business when legal risks soared led to a government shield to protect industry from some liability. Contraceptives perform a similar social function and produce a similar social good. Their manufacturers should be afforded similar legal protection. Finally, the report urged the development of a private–public partnership between industry and the federal government (NIH and USAID) for the future development of new contraceptive methods. ■

Industrial R&D Decision-Making and Contraceptive Research

Based on remarks delivered to the Science and Technology Policy Forum of the New York Academy of Sciences on April 15, 1997.

MICHAEL E. KAFRISSEN
Vice President for Clinical Affairs
Ortho Pharmaceuticals

P rivate industry possesses important capacities in the development of new approaches to contraceptive technology. It is important, however, to begin by understanding how pharmaceutical companies make research and development decisions.

No Lack of Choices

Humanity does not lack for health problems that need solutions and any pharmaceutical company has many therapeutic candidates that compete for limited R&D resources. For private industry, the

> **Private industry spends some $19 billion in research on ethical pharmaceuticals every year.**

responsibility for responding to public health needs is complemented by the responsibility to shareholders. There are obviously choices to be made; no company can do everything it would like. To make those necessary choices one must ask what resources are available, what costs will be incurred, and what organizational and market elements will affect the likelihood of success.

Resources

Currently, the U.S. ethical pharmaceutical industry invests $18–19 billion in research and development. In terms of corporate resources, this represents 21% of sales. Relative to other industries this is a fairly high commitment of available capital to research. Clearly, the pharmaceutical industry is a player in the health research field.

Costs

But however much is spent, how much does successful research cost? The U.S. Pharmaceutical Research and Manufacturers Association estimates that it costs in excess of half a billion dollars per new chemical entity for an innovation to reach market. Hence, new drugs are huge investments. The total research budget, representing a significant portion of any company's resources, must be carefully managed if it is to be spread effectively over such a cost structure.

It is true that technology is enabling much more efficient research. Screening and testing of components for potential utility, for example, is much quicker and less costly with high throughput technology. Nevertheless, even if a company is

offered partnership in producing a
new product that is already going
into phase two trials, the expenditure
stream before it reaches market is still
extremely daunting.

> **In the pharmaceutical industry, 21% of sales are reinvested in research. Recent estimates suggest that it takes in excess of half a billion dollars to bring a new chemical entity to market.**

Decision Elements

In addition to the simple level of
costs, there are a series of considerations that guide corporate R&D decision making.

1. Therapeutic Need. Unmet or inadequately met patient needs
predict demand. The magnitude and severity of the condition
influence the interrelated public health and market assessment.

2. Geography. The United States represents the largest portion
of company-financed R&D as well as the largest market. Not sur-
prisingly, western Europe and Japan are the other major players.
Research opportunities that relate to these markets, then, are
understandably attractive.

3. Attrition. Thousands of entities are screened and tested for
every one that yields enough potential to enter into clinical trials.
Many products that make it into such trials never fulfill their hoped
potential. So there is a tremendous attrition of chemical entities as
they progress through the testing and approval process. In fact, over
the years, the financial burden associated with clinical trials in the
United States has increased. The burden is not only in terms of
direct costs, but also in terms of time and human resources. A current New Drug Application (NDA) filing with the Food and Drug
Administration can total 90,000 pages of information.

4. Competition. For industry, current and future competition is
a critical decision element. An anti-progestin may be entirely fasci-

> A full filing for approval of a new drug entity at the Food and Drug Administration can require 90,000 pages of supporting documentation.

nating in the research laboratory, but if another company is soon to make an anti-progestin available, it may affect the investment decision.

5. Corporate Fit. How long will the drug take to manufacture? How expensive will it be to sell? Does the company have the types of skills necessary in its human resources to market and sell such a drug? The "fit" to a company's existing line of therapeutics is an often elusive but nonetheless important consideration in making R&D investment choices. For a company that has never been in the gastro-intestinal business to take advantage of suddenly acquiring a gastro-intestinal drug will require huge investments in human resources and organizational networks. This is why it becomes dangerous if only a few companies remain in a particular product (for example, contraceptives); the possibility that the therapeutic offerings will narrow and not grow is very real.

6. Patents. Patent protection is essential, both in terms of length of time that patent protection will be granted, and the likelihood that patents will be respected in the drug's key markets.

7. Line Extension Potential. Companies are particularly attracted to opportunities that address multiple needs. An angiogenesis antagonist, for example, may stop blood vessel growth in malignancies, but it may also hold potential for application to retinal or endometrial diseases. Given the size of the investment that must be made in any successful research effort, leverage is attractive. Potential application of one drug to other needs increases the overall return on the research investment.

8. Liability. In this litigious society, exposure to liability is a major consideration for any pharmaceutical company. Some companies have seen significant financial loss from liability controversies over contraceptives; others have been bankrupted by the law suits. The industry is very aware of the legal dangers in this product line, and that awareness

> **We can do a better job of aligning public health needs and private sector product decision-making**

raised significant hesitations about becoming involved in the contraceptive business. Risk–return analysis must be considered.

In Summary

It is certainly possible to do a better job of aligning public health needs with private sector product development decision-making. The private sector is market driven; corporations are eager for new opportunities, particularly if they are consistent with their existing products and capacities. What is truly needed is a broad public-private consideration of priorities in health care technologies that would yield both a qualitative and a quantitative strategic planning tool to identify opportunities in contraceptive techniques and markets in the context of private sector decision-making. The joint demands of the public health and the private sector are not incompatible. Rather it is a necessity for formulate partnerships for successful therapeutic development. ■

A Clear and Present Threat to National Security: Military and Civilian Vulnerability to Biological Weapons

Based on remarks delivered to the Science and Technology Policy Forum of the New York Academy of Sciences on October 30, 1996.

JOSHUA LEDERBERG

University Professor, The Rockefeller University

D isease has been associated with warfare and expeditionary forces since time immemorial. History has been grimly shaped, for example, by the inadvertent movement of smallpox and measles across the Atlantic from the Old World to the New World, and, some would believe, the transmittal of syphilis in the other direction. During Pacific campaigns in World War II, malaria claimed as many casualties as Japanese bullets.

Biological Weapons and Military Capacity

However, since the British use of smallpox-infected blankets during the French and Indian Wars, there have been no major incidents of intentional, malicious biological warfare in actual military practice. This fact should provide little comfort. Intelligence about German experiments in biological warfare development during World War II sparked a major U.S. effort at Fort Dietrich. During the Cold War, a multi-polar arms race, with the U.S., Britain, and the Soviet Union as the major players, included biological warfare components. The intent of the 1972 Biological Weapons Convention, the first categorical disarmament in recent history, was to end the development of these types of weapons of mass destruction. The race, we thought, was over, the victory won.

It was, therefore, a great disappointment, although perhaps not a great surprise, to learn in 1979 that despite signing the treaty, the Soviet Union had continued a clandestine military biological weapons program that resulted in an unprecedented outbreak of anthrax directly as a result of an accident at a military installation. Mercifully, the United States and its allies have totally dismantled their offensive biological weapons programs. We know, however, that Iraq and many other rogue states have not. These nations have sustained their offensive programs and, in many cases, have weaponized biological warfare agents so that they are ready for delivery at a moment's notice.

The Civilian Threat

There is growing recognition among military leaders that conflict with these nations could result in the need to defend against biological agents. Yet, it is terrorists and small, violent groups for whom these weapons hold great attraction. It is civilian populations, therefore, that are really vulnerable and therefore attractive targets. The United States has hardly begun to organize itself for civil defense against these weapons.

Why are civilian populations so vulnerable? In part this is a function of what it takes to make the biological agents, and in part what it takes to deliver them.

The defining distinction of biological weapons is that they use live organisms as a source of disease in order to initiate a continued infection. That infection depends on the proliferation of the

> Since the British use of smallpox-infected blankets during the French and Indian Wars, there have been no major incidents of intentional, malicious biological warfare in actual military practice. This fact should provide little comfort.

bacteria or the virus once it has entered into the body. The catalogue of possible infectious agents is large, and it is not secret.

A Case in Point: Anthrax

The common thread through most of the agents is bacillus anthraces, the anthrax organism. Anthrax seems in many ways to have been invented by nature almost uniquely for biological warfare, and it represents the greatest risk.

The properties of anthrax provide it with tremendous magnitude as a weapon. It is more potent, more readily available, has a long shelf-life because it can produce highly resistant spores, and can be transmitted by aerosol through the atmosphere. It is also cheap to produce. No other agent has all of these properties.

What would be the effect of delivery of an anthrax-based weapon? To quote from the Office of Technology Assessment: "A hundred kilograms of anthrax spores optimally disseminated under ideal meteorological conditions, would have about the same potency in civilian casualties as a nuclear weapon of comparable size, but would be a thousand to a million times cheaper to produce and more readily accessible to a wide range of actors."

What would it take to obtain such a weapon? Production could be done in a room the size of the Main Hall of the New York Academy of Sciences (20 feet by 40 feet). Of course, it would take time to produce enough material—perhaps a few weeks. Of course, time might not be a problem, since it would be difficult to determine that a weapon was being developed. Once enough spores had been produced, the "delivery vehicle" might be an aerosol container sprayed from the top of the Empire State Building. Add in a clear day and a pleasant breeze, and a substantial portion of the population of Manhattan would be at risk.

What Can Be Done?

A central problem, of course, is that the threat of biological weapons will always exist, and, ironically, as our scientific understanding expands, so does the potential for biological weapons. Studies in microbiology are inherently of dual use. They are primarily motivated by the necessity of understanding infection and developing defenses. But as soon as that knowledge is obtained, it is instantly available for malicious application. This technical base for biological weapons is the very technical base we must rely on for advance in biological sciences. A pharmaceutical production facility being used today for innocent, indeed, life-saving purposes, could be transformed in a few weeks or months into a production facility for any of these agents. The problems of control and authentication obviously are complex. Nowhere are intention and capability more confusedly intertwined than in biological weapons.

> **The United States has hardly begun to organize itself for civil defense against these weapons.**

There is good news of sorts. In a biological event, casualties are substantially delayed. Once an individual has been exposed to a nuclear blast, not much can be done, if there is anything left, to alter the outcome. However, the fate of individuals exposed to biological infectious agents can be modified by enormous factors. With appropriate medical intervention, ninety-five percent casualty rates can be reduced to one or two percent. But that result will only be obtained with significant advanced preparation.

> **Anthrax seems to to have been invented by nature almost uniquely for biological warfare. And a hundred kilograms could be produced in a room the size of the Main Hall of the New York Academy of Sciences.**

The Second Edge of the Sword

Biological weapons represent a looming national security threat. Nothing less than the possibility of civil life is at stake if these kinds of weapons become widely available and are used. But the nature of these weapons raises a real hurdle to developing treaty mechanisms for their prevention and control. Biological weapons are a two-edged sword, and the second edge is poised at the heart of civil structure in a free society.

> **Studies in microbiology are primarily motivated by a desire to understand infection and develop defenses. But, as soon as that knowledge is obtained, it is instantly available for malicious application.**

Biological weapons represent a threat to civil life not merely in their use but in the means needed to prevent their development. Preventing the possibility of biological weapons development in otherwise innocent places would require a degree of intrusive oversight of the activities of every individual that would be incompatible with our definitions of freedom.

> **We must build up an international community of resolve that the use of biological weapons will simply not be tolerated. Not now, not ever.**

Confidently preventing any individual from developing the means to destroy society would require the possibility of intrusion into the life of every individual. If an anthrax weapon could be developed undetected in the Academy's Main Hall, then every room in every institution of even a comparable size would be the potential target of unannounced visits by weapons verification teams.

The weapons themselves are unacceptable, and the means for assuredly preventing their development are equally unpalatable. What is to be done?

We must build up an international community of resolve that the use of biological weapons will simply not be tolerated. There is too much at stake to do otherwise. In many, many situations, enduring coalitions have escaped humanity's grasp. In the case of biological weapons, however, there really is no other choice. An international coalition of shared resolve is the only defense in the long run. It will require a global partnership of the scientific, medical, and business communities, exerting leverage on the political processes of nations, to create a clear, unambiguous message: humanity will not accept the presence of biological weapons on the planet. Not now, not ever. ∎

Responding to the Biological Warfare Threat: Possible Approaches and Necessary Capacities

Based on remarks delivered to the Science and Technology Policy Forum of the New York Academy of Sciences on October 30, 1996, in his capacity as United States Under Secretary of the Navy.

RICHARD DANZIG

United States Under Secretary of the Navy

W hy has it been extremely difficult to get people to attend to the threat of biological warfare? Leaders in science, such as Nobel Laureate Joshua Lederberg, have been warning for years that this is a national security threat. This problem confronts the military. This problem confronts civilian society. This problem could be our undoing. This problem is real and present. Informed leaders agree.

Equally, we know that some terrorist groups have the capability to produce these weapons. A substantial number of nations have, in fact, active biological weapons programs. We know that the ability to weaponize anthrax is out there. We know what a kilogram of anthrax will do.

Why is the threat of biological warfare so low on the agenda of

> **Why is the threat of biological warfare so low on the agenda of action as to be virtually invisible?**

action as to be virtually invisible? Three propositions are generally offered as the rationale for failure to focus and act. Each of the three is misguided.

Proposition Number One: A Nuclear Response Will Suffice

If biological weapons can be sized to small group use, and if they can be developed in small settings, and if they are relatively cheap, then such weaponry can be generated without nation states. Moreover, these weapons involve a delay of onset of symptoms, perhaps as much as 24 hours, rendering low visibility to the actual incidence of their use.

It is not apparent that nuclear weapons can be used to stop terrorist groups. It would be even more difficult to target a nuclear response against a small group which is no longer present at the point of a biological weapons incident at the time that the incident is discovered.

Justifying a nuclear response is also complicated by the fact that the biological entities at issue exist in nature. It could take time to determine that a disease outbreak was a result of an intentional, hostile act rather than a natural occurrence, further complicating the judgment regarding retaliation and distancing the perpetrators from the use of the weapon. Moreover, any nuclear response, in light of all of these shades of grey and possible uncertainties, would open the United States up to tremendous international criticism.

A nuclear response to a biological weapons event is a snare and a delusion.

Proposition Number Two: Biological Weapons Will Never Be Used

The second rationale for not taking serious steps to protect against

biological weapons is the view that
they will never be used. There is a
sense that they are so immoral that
their use would generate a level of
public revulsion that itself acts as a
deterrent. This view points to the fact

> **A nuclear response to
> a biological weapons
> event is a snare and
> a delusion.**

that they have never been used as evidence that they will never be
used.

The flaw in this argument lies with experience. No significant
weaponry has been developed that has not been used. Weapons do
not remain theoretical; when they are available, they are used. The
probabilities that biological weapons will forever defy past experience are not high.

Furthermore, history teaches that biological weapons have been
used. In the Middle Ages, attacking forces catapulted cadavers over
the walls of cities under siege to create outbreaks of disease. The
British infected blankets with smallpox during the French and
Indian War. In the U.S. Civil War, when General William
Tecumseh Sherman was embarked on his famous march to the sea,
Southern defenders poisoned the local wells to slow his advance.
This decade, a terrorist group in Japan made efforts to develop biological weaponry; chemical alternatives were simply faster and easier. So there is no necessarily huge psychological barrier that would
inhibit nations or groups, who perceive themselves deeply embat-
tled, adding biological weapons to the mix of their active arma-
ments.

Proposition Number Three: Non-Nuclear Action Against Biological Weapons Is Too Difficult

Biology is not familiar terrain to the military. World War I created
in the military an appreciation of the importance of chemistry.

> History teaches that biological weapons have been used. There is no necessarily huge psychological barrier to inhibit those who perceive themselves deeply embattled from adding biological weapons to the mix of their armaments.

World War II saw a similar expansion of the relationships between the military and the global community of physicists. The Cold War witnessed deeper linkages between military thinking and capacity and the evolution of innovation in the telecommunications and electronics sciences. Biology has not experienced a similar relationship with military leadership. Hence, within the military there is not a great understanding of biological strategies and options in confronting this threat.

Five Approaches and Needs

But science and technology do, in fact, provide us with very real options in defense against biological weaponry. Five examples will serve to illustrate the degree to which action against biological weapons is feasible, yet requires changes in the way the nation thinks about and organizes for a confrontation.

First, biological attack is essentially invisible. Response requires warning. Detector technology is advancing rapidly, and the United States is now beginning to field detector capabilities. Much more investment in this capacity is required.

Inoculation is an obvious additional example. Vaccines exist for a number of biological threats, including anthrax. While they have been used in the face of an immediate state of emergency, including in the Gulf War, they have not been routinely administered to military troops. They ought to be made part of troop preparation whenever the military is in a vulnerable setting. In addition, however, increased investment in research is needed, both for vaccines and for antibiotics. This is an area in which there must be very close collaboration between the United States Government and the private pharmaceutical industry.

> **Innoculation should be made part of troop preparation whenever the military is in a vulnerable setting.**

A third area of progress, and one in which further investment is mandatory, is intelligence. Finding a production facility for biological weaponry that could be a back room in any building in the world poses obvious challenges to the intelligence community. Overhead satellite imagery is a critical capability when you are looking for missiles and nuclear production facilities; it is of little utility in finding that back room. The analogous experience that is available, however, is that of drug enforcement authorities. The military would do well to work closely with national drug monitoring and interdiction agencies to determine technologies and approaches that might be applied to biological weapons intelligence gathering.

Fourth, there is a need to forge a closer working relationship between civilian and military authorities. The job of the military is to defend the nation. To date, that defense has predominantly been against foreign enemies fighting abroad. That traditional distinction between military defense abroad and civilian law enforcement at home beaks down in the face of terrorism. If biological weapons could pose a terrorist threat to American civilians, then all of the nation's resources must be coordinated to prevent or respond to such an eventuality. If there was a biological attack on New York, it

> **Developing effective tactics against biological weapons will require close collaboration between the military and the scientific community.**

would not be a problem simply for the New York City Police Department. A coordinated approach must be forged, in which civilian agencies, such as the FBI, the public health service, and police departments, have an integrated plan that includes military capacity. Lines of communication and procedures for action must be established between civilian and military authorities. Fragmented capabilities in the face of a biological weapons incident will have tragic consequences.

Fifth, we need entirely new ways of thinking about military tactics in the face of biological weapons. Biologicals are non-explosive, invisible, insidious weapons. There must be a dedicated focus of military thinking on how to defend against and respond to biological attacks. In turn, this will require closer collaboration between the scientific community and the military; biologists will need to work closely with military strategists to identify the most effective approaches to these types of weapons. ■

4

Industrial
Dimensions

The biotechnology industry in the United States constitutes some 1,300 companies representing $13 billion in sales. The industry's $9 billion investment in R&D represents a 58% growth between 1993 and 1997. On average, biotechnology products require 10 years and in excess of $250 million to reach market. The industry projects sales of $20 billion by the year 2000, reaching markets in medicine and agriculture around the world.

In the New York region, the combination of world-class academic institutions and industrial leadership in pharmaceuticals and medical devices has created significant opportunities for exploiting biotechnology discoveries and applications. New York State ranks first in the nation in terms of the number of institutions of higher education, and fifth in terms of the number of biotechnology companies located in the state. Nearly two-thirds of these companies are focused on medical diagnostics and therapeutics markets.

Yet, although New York's technology roots are deep and its assets are significant, there is growing unease about the region's overall technological health and consequent implications for the economy. The New York region's share of national patents dropped from 20 percent in the mid-1970s to 16 percent in 1994. Such trends pose long-term concerns for many technology-based industries, including biotechnology. The share of employed Ph.D.s and engineers in R&D is also falling. Continued investment in R&D capacity and skilled human resources is critical to biotechnology and all high-technology industries.

The volume and depth of research is crucial because research support seeds the scientific discovery upon which product development is based. The support of the National Institutes of Health, for example, remains a crucial anchor for American life science and for U.S. competitiveness in biotechnology. But New York lags behind other regions of the country in attracting NIH grants.

This paper is the result of a three-session colloquium jointly convened in the fall of 1995 by the New York Academy of Sciences and New York University. Funding for the series was provided by the Carnegie Corporation of New York. The first session examined the foundation of molecular biology upon which biotechnology is based, with discussion of associated economic and social implications. The second session addressed the ways in which advanced research in universities is linked to private capacity, and discussed how these partnerships might fully harness the potential of biotechnology. The third session examined the status and prospects of the biotechnology industry in the New York region.

This paper presents a summary of the remarks and comments made at the symposia, with updated statistics reflecting recent trends in the biotechnology industry. Except where attribution is explicit, the report is not a verbatim record of proceedings. The report, which was written by Alexandra Levitt, a consultant to the Academy's policy staff, was derived from speakers' presentations (see the program at the back of this report): Thus, chapter one, "Molecular Biology and Society," is derived from presentations by Joshua Lederberg, Richard Axel, Nam-Hai Chua, and Dorothy Nelkin; chapter two, "Commercialization of Biotechnology Innovations," is derived from presentations by Christine Dietzel, Nam-Hai Chua, Joan S. Brugge, Sandra Panem, Roger C. Herdman, and Rochelle C. Dreyfuss; and chapter three, "Biotechnology in New York," is derived from presentations by Margaret A. Hamburg, Jack Huttner, Herbert Pardes, and Lennart Philipson.

Industrial Dimensions

Sometimes perspective helps. The biomedical industry traces its roots deeply into human history. The bright future of the industry is perhaps made richer in texture by a brief, selective parting of the mists of time.

Egyptian carvings show surgical operations in progress	2500 BC	
	2100 BC	Oldest medical text written on cuneiform tablets
The Papyrus Ebers provides a description of 700 medications	1550 BC	
	40 AD	*De materia medica* by Greek physician Pedanius Dioscorides of Anazarbus deals with medical properties of about 600 plants and nearly 1000 drugs
Trotula of Salerno advocates cleanliness, a balanced diet, exercise and an avoidance of stress as keys to good health	1070 AD	
	1240 AD	By decree, the Holy Roman Empire permits the dissection of human cadavers
Roger Bacon discusses spectacles for farsightedness in *Opus majus*	1268 AD	
	1284 AD	Mamluk Sultan Qalawin builds Mansuri Maristan in Cairo, the most sophisticated medical center of its time
The Black Death reaches northern Europe	1352 AD	
	1473 AD	The first complete edition of Avicenna's *Canon of Medicine* is printed in Milan

Source: Bryan Bunch and Alexander Hellemans, *The Timetables of Technology; A Chronology of the Most Important People and Events in the History of Technology.* (New York: Simon and Schuster, 1993.)

The Royal College of
Physicians is established — 1518

1535 — *Dispensatorium* by Valerius
Cordus describing most known
drugs, chemicals and medical
preparations is published

Robert Boyle demonstrates
ability to keep animals alive by
artificial respiration — 1667

1701 — Giacomo Pylarini innoculates
three children with smallpox in
hopes of preventing more serious cases when they are older

Giovanni Morgagni publishes
the first important work on
pathological anatomy — 1761

1775 — William Withering introduces
digitalis for dropsy associated
with heart disease

Edward Jenner gives first
inoculation using cowpox to
prevent smallpox — 1796

1797 — The Royal Society rejects
Jenner's technique

Friedrich Gustav Jakob Henle
expresses his conviction that
diseases are transmitted by
living organisms — 1840

1865 — Joseph Baron Lister reduces
surgical death rate from 45 to
15 percent after using phenol
as disinfectant

Robert Koch discovers that the
microorganism responsible for cat-
tle anthrax can be grown in culture — 1876

1881 — Louis Pasteur develops the first
artificially produced vaccine

Felix Hoffmann synthesizes
asperin — 1893

1905 — George Washington Crile per-
forms the first direct blood
transfusion

Rose G. Harrison demon-
strates the *in vitro* growth of
living animal tissue — 1907

Ernest Goodpasture grows
viruses in eggs

1931

1929 Alexander Fleming discovers
penicillin

Selman Abraham
Waksman coins the term
antibiotics; discovers strep-
tomycin in 1943

1941

1935 Gerhard Domagk uses the first sulfa
drug on his youngest daughter to prevent her death from streptococcal
infection, the first use in humans

Mass polio inoculations
begin in U.S. using the
Salk vaccine

1954

1953 Watson and Crick

1963 First lung transplant by James
Daniel Hardy

First successful heart transplant by Christiaan
Neething Barnard

1967

1975 Cesar Milstein and Georges J.F.
Kohler announce discovery of how
to produce monoclonal antibodies

Genetic code for the hepatitis
B surface antigen is found

1981

1986 U.S. Food and Drug
Administration first approves
monoclonal antibodies for therapeutic use in humans

Birth of Dolly is
announced to the world

1997

1998 Total market capitalization of
biotech companies in the U.S.
tops $93 billion

Molecular Biology and Society

olecular biology is a young science. Its genesis, however, spans many scientific landmarks:

- One hundred and thirty-two years ago, in 1866, Gregor Mendel experimented with pea plants and theorized about what a unit of heredity, a "gene," might be.

- Eighty-eight years ago, in 1810, Thomas Hunt Morgan at Columbia University, demonstrated that genes lie in linear arrays along chromosomes.

- Fifty-four years ago, in 1944, Oswald Avery at Rockefeller University discovered, with Colin MacLeod and Maclyn McCarty, that DNA is the chemical substance of the "transforming principle," i.e., DNA carries genetic information and can cause heritable change in bacterial cells.

- Forty-five years ago, in 1953, James Watson and Francis Crick elucidated the structure of DNA, and scientists began to decipher the genetic code.

Biotechnology—the industry that grew out of molecular biology—is younger still. Less than twenty-five years ago, in the mid-1970s, scientists discovered that fragments of DNA from different sources (viral, bacterial, plant, or animal) could be cut apart and spliced back together in different combinations, using bacterial enzymes called restriction enzymes. The recombined or "recombinant" DNA remained perfectly functional when inserted into a bacterial cell.

Milestones in Biotechnology

Scientific and legal landmarks in the history of biotechnology include:

- The discovery of restriction enzymes
- The first recombinant DNA experiments, in which genes encoding SV40 tumor virus proteins were spliced into bacterial plasmids and expressed in E. coli
- A historic meeting of molecular biologists at Asilomar, California, in 1975, to consider the potential hazards of the new technology
- The Supreme Court Decision that allowed engineered life forms to be patented

DNA: Science and Technology

Over the last 50 years, science has made enormous progress in the analytical understanding of DNA. The study of microbial, animal, plant, and human genetic material has provided far-reaching insights into health and disease and may ultimately lead to insights into human "nature" itself. In addition, improved understanding of DNA as a chemical entity has spawned a new generation of technologies and industrial applications.

Science now possesses the tools to isolate DNA efficiently and to amplify single DNA molecules in large quantities. The scientific "workshop" also includes an impressive array of "seamstress tools"—enzymes that alter DNA—creating the capacity to stitch, splice, and tailor DNA sequences to fit different uses. These tools allow the study and manipulation of genomes of simple as well as complex organisms.

Scientists can move gene fragments from the genome of one organism to another and insert them into particular chromosomal sites where they will be expressed. DNA fragments containing human genes can be "shot-gunned" into bacteria, so that each microbe carries a single fragment. Because bacteria multiply a million-fold in a very short time, scientists can use specific probes and selection strategies to "pull needles out of haystacks"—that is, to isolate a one-in-a-million bacterial clone that carries a gene encoding a specific human protein. Such technical tools have been very useful for discovering genes in humans and other organisms.

Advances in Biotechnology

In turn, these advances have generated a wide array of applications.

Human Drugs. Biotechnology is now a multi-billion-dollar industry, with pharmaceutical biotechnology companies enjoying the largest financial pay-off. Because drugs like human tissue plasminogen activator (TPA), which is used to lessen heart muscle damage following a heart-attack, have dramatic, life-saving properties, it is sometimes possible to recover the substantial cost of developing and testing a product. But biotechnology promises no immediate industrial bonanza for human drugs. A few big successes have been accompanied by many failures.

Transgenic Plants and Animals. Genetic engineering in plants and animals has also become technically easier. The basic principles are the same as in microbes, although it is not possible to apply million-fold selection schemes. Transgenic plants, including tomatoes, corn, cotton, and soybeans that are resistant to insects or herbicides, are becoming commercially available. Their use will allow farmers to apply fewer chemicals in their fields. In addition, transgenic animals are currently used in basic research as models of human disease or to investigate the function of human or animal genes.

In the future, transgenic farm animals may be used as "living factories" that produce proteins for pharmaceutical or industrial use. Instead of manufacturing a protein—such as an immunoglobulin—in a fermenter, the biotechnologist may breed a transgenic cow that can secrete the protein into her milk. In some cases it may prove easier and cheaper to use a fruit or vegetable "factory" (a banana or potato) rather than an animal.

Reproductive Technologies. DNA technology is a core part of reproductive technologies now used to generate transgenic animals and plants. In principle, techniques used with a mouse embryo could also be used with a human embryo; there does not seem to be any fundamental scientific issue in human embryonic development that distinguishes it from the equivalent process in mice. Obviously, the traffic light at this intersection of science and society flashes bright red.

The Human Genome Project

Perhaps nowhere is the impact of the tools of biotechnology more evident than in the project to map the entire human genome. The project has led to information about the sequence of approximately three-quarters of the genes on human chromosomes. However, despite this rapid progress, the major lesson learned from the Human Genome Project is how much we still do not know. Of the thousands of human genes mapped so far, biologists can assign a probable function to only about one-third.

Polymorphisms and Individual Differences. The Human

Genome Project has shown us how polymorphic we are, i.e., how much genetic variation there is between individuals. There is a large degree of heterogeneity in the human population, and it is clear that the significance of racial differences in that heterogeneity has been greatly exaggerated. Beyond the most superficial differences, e.g., skin or hair color, the amount of heterogeneity between races is much smaller than the amount of heterogeneity between individuals of the same race.

Humans and Other Primates. Data from the Human Genome Project confirm that the human genome is very closely related to the genomes of other primates. In fact, the same DNA polymorphisms are found in individual apes and in humans, indicating that the polymorphisms occurred before the human line diverged from the apes.

Some Practical Applications of Molecular Biology

Forensics. DNA fingerprinting is now widely used in forensics to identify criminals. The FBI is planning to maintain a registry of genetic material from rapists and violent criminals. Since many violent criminals are repeat offenders, it is hoped that the DNA registry will prove useful in criminal investigations. It should be kept in mind, however, that these forensic applications raise issues of civil rights

Fetal or Embryonic Diagnostics. DNA technology is well advanced in the area of prenatal diagnosis of genetic diseases. It is theoretically possible to eliminate a disease like Tay-Sachs—whose victims die by the age of three—by "pre-empting" embryos before they become fetuses. At the present time, diagnostic procedures such as amniocentesis are performed during the first or second trimester of pregnancy. The present procedure is to use blood tests to screen prospective parents to determine whether they carry the recessive, disease-causing form of the Tay-Sachs gene. If both parents carry the recessive gene, diagnostic procedures are performed to examine the fetal chromosomes. If, however, the fetus carries a pair of mutated genes, science and technology again impact on societal (or in this case, individual) decision-making. Science can elucidate the facts, but it cannot make the decisions.

Confidentiality

There is growing concern that information about an individual's genetic predisposition to specific diseases, taken from medical records and the results of genetic testing, could be abused if made widely available. Negative individual consequences could be caused by the reactions of, for example, insurers and employers. On the other hand, a total information bar is not necessarily advisable. Genetic information revealed to the person it came from—and the person's family and doctor—might be of great value in medical and personal decision-making.

> In regard to life insurance, if a person has exclusive access to his or her own genetic information, then it changes from a crap game to a poker game.
>
> **Joshua Lederberg**
> **The Rockefeller University**

Future Medical Interventions

Advances in molecular biology have led scientists to investigate new types of medical interventions that affect health by altering gene expression. Depending on the methodology, the effects of the proposed interventions may be transient, or they may endure for the lifetime of the individual. Or, the effects may be passed on to an individual's offspring.

One methodology that promises to alter gene expression on a temporary basis is "anti-sense" therapy, which aims to prevent the expression of particular proteins that cause or exacerbate disease. This may be accomplished using artificially synthesized DNA or RNA "anti-sense" molecules that bind to targeted genes or messenger RNAs that carry complementary ("sense") sequences. Because the genes or messenger RNAs are "tied up" by the anti-sense molecules, the proteins encoded by the genetic material cannot be expressed.

Most methodologies that attempt to alter gene expression permanently come under the heading of "somatic cell therapies." These interventions aim to repair a genetic defect by replacing a missing or mutated DNA sequence, or by deleting a deleterious gene. In theory, somatic cell therapy could be used to stop the growth of a cancerous cell, or to enable to immune system to manufacture antibodies against specific pathogens. Although techniques for introducing DNA into a cell are already available, problems remain on how to target a new gene to a particular chromosomal site.

The most controversial methodologies are "germline therapies," which alter genes in sperm or eggs (or in their precursor cells) and affect the next generation. Although the technical procedures involved in somatic and germline therapies are quite similar, their effect is hugely different. Deliberate alteration of the germline— even to eliminate a genetic disease—raises serious ethical questions that society, including scientists, must confront.

New Frontiers in Molecular Biology

Throughout the 1980s and 1990s there was a continuous flow of new discoveries in human molecular biology as scientists applied molecular techniques to one new area after another. During the past year, human genes involved in the etiology of breast cancer, colon cancer, and obesity were identified and isolated. In addition, biologists detected gene products that inhibit angiogenesis (the branching-out of oxygen-supplying capillaries around cancer cells), a discovery that may revolutionize the treatment of metastatic cancers. Moreover, DNA sequencers produced extensive information on bacteria that cause human diseases, including ulcers and tuberculosis. Finally, there are promising new approaches to the study of nerve regeneration, aging, and the immune system.

Two different and exciting areas of molecular biology are illustrated briefly below: the molecular biology of plants (which has seen an explosion of discoveries over the last few years) and the molecular biology of human perception (which is just beginning).

The Molecular Biology of Plants

Several technical advances have made it possible to bioengineer new strains of transgenic plants. These advances have led rapidly to many commercial applications, and many more are expected within the next few years.

Agricultural Benefits of Transgenic Plants. Transgenic food plants (e.g., corn, rice, soybeans, tomatoes, potatoes, and wheat) that can grow in harsh environments or have increased nutritional value may enhance the world's food supply and benefit people all over the world. In addition, transgenic plants that are genetically resistant to bacteria, viruses, or fungi will reduce the load of chemicals in the environment, because farmers will rely less heavily on pesticides.

In the past, plant biologists feared that the public might not accept genetically engineered foods. However, that does not appear to be the case, at least in the United States, where the "Flavr Savr" tomato is selling well to consumers.

Basic Research in Plant Biology. Continuing progress in plant biology requires a deeper understanding of plant genes and how they are regulated. Molecular biologists seek to isolate new genes in different plant and non-plant species; to increase their understanding of plant development; and to understand how cells communicate at the molecular level (signal transduction). As scientists investigate these basic questions, they inevitably make discoveries about the chemistry of living things. Thus, the study of plant biology at the molecular level is having a dual impact. It is changing our ideas about plant life and evolution, and it is starting to have a profound effect on the agricultural industry.

The Molecular Biology of Perception

DNA is the genetic material of all identified species on earth. It consists of the same molecular building blocks in such diverse entities as microbes, trees, and humans. In addition, the cells of all known organisms share many molecular structures and processes that can be examined by molecular methods. Even the cells that make up the brain—the neurons—can be studied at the molecular level like liver cells or bone cells. Thus, fundamental questions of neurobiology can be investigated by molecular techniques.

A Case in Point: The Sense of Smell. The sense of smell is the primary sensory modality for most animals, allowing them to recognize predators, food, and mates. The olfactory neurons constitute an evolutionarily primitive sensory system that can be used as a model to study perception.

In human beings, there are two distinct sets of olfactory neurons in the nose. Each set contains 1–5 million neurons, each of which sends a very long axon (a tail-like projection of the cell) back to the brain. Thus, there is a direct, monosynaptic connection between the olfactory neurons (the sensory receptors) and the brain.

One set of neurons represents the "main nose," which mediates the conscious perception of pleasurable and aversive odors. The axons of this group of neurons form synapses with cells in the brain's cerebral cortex, the site of conscious actions and higher reasoning. The second set of neurons —sometimes called the "erotic nose"—mediates the unconscious perception of odors that provide clues to the social and sexual status of individuals. It affects such behaviors as mothering, mating, and fighting. The axons of the "erotic nose" bypass the cognitive cortex and form synapses with neurons in the amygdala and hypothalamus, parts of the brain that are concerned with emotions.

How Does the Nose Recognize Specific Odors? Until recently, very little was known about how we recognize and distinguish tens of thousands of different odors. Do the nasal neurons contain a small number of general receptor proteins, each of which reacts with many different smell-inducing molecules? Or do they contain a tremendous number of specific receptors, each of which detects a different odor?

Molecular biologists have addressed these questions by cloning the genes that code for the proteins that serve as odorant receptors. They found that the receptor proteins lie within the neuron's outer membrane—weaving in and out of the membrane seven times. They also found that there are at least 1,000 genes in the human genome that encode odorant receptors—the largest gene family yet discovered in mammals. Evidently, odor recognition involves a vast array of genes for different odorant receptors.

How Does the Brain Know What the Nose is Smelling? How does the brain know which nasal receptor has been activated by a particular odor? It turns out that each neuron makes only one type of receptor protein. Therefore, the question becomes: How does the brain know which neuron has been activated by a particular odor? The answer is that each neuron forms synapses with one discrete locus in the brain. That is, there is a one-to-one correspondence between the projections of a single neuronal axon and a specific site in the brain. Out of these sites, the brain makes a two-dimensional olfactory map. This map is apparently identical in all individuals of a given species.

Since there are 1,000 receptor proteins and humans recognize about 10,000 odors, the recognition system must be based on combinatorial sets of inputs. That is, a particular odor (such as roses) must activate not one, but a defined set of receptors that activates a defined set of sites on the brain's olfactory map. If there are 1,000 receptors and—perhaps—10 loci in a combinatorial set, that means that there are millions upon millions of possible combinatorial patterns. Since humans cannot in fact distinguish among such a huge number of odors, scientists infer that the brain selects for attention only those combinations that are important for the survival or reproduction of the species. Those combinations comprise the set of odors that human beings can smell.

This model of the olfactory system is similar to the current model of the optical system. In the case of vision, the sensory image is apparently dissected into distinct components—color, orientation, size, distance, motion—that are detected by specific sets of receptors. The components are then communicated to the brain as individual, parallel inputs that produce combinatorial patterns on the visual brain map.

<u>Where Molecular Biology Leaves Off</u>. Thus, science now has molecular evidence for a precise two-dimensional map that provides the brain's "pre-knowledge"—or organizing principle—for olfactory perception. However, it may not be possible to answer a fundamental question—*Who or what interprets the image in the brain?*—by designing a molecular biology experiment. The answer may have to await a new and transforming insight into the nature of human consciousness.

DNA in Popular Culture

American popular culture integrates scientific ideas in ways that reflect social tensions and political agendas. Popular notions about genetics—as reflected in magazines, talk shows, ads, sitcoms, and soap operas—often define "normal" relationships, set policy agendas, and frame ideas about appropriate behavior.

Dorothy Nelkin and Susan Lindoe, the authors of *The DNA Mystique: The Gene as A Cultural Icon,* coined the term "genetic essentialism" to describe the popular belief that human beings in all their complexity can be "reduced to DNA." Genetic essentialism includes three assumptions that are pervasive in the United States:

- Genes are the critical essence of personal identity and social bonds.
- Genes determine behavior.
- Future health and behavior can be predicted through genetic analysis.

All three beliefs have profound social consequences.

Genes as the Essence of Personal Identity

The first belief—that genes are the critical essence of personal identity and social bonding—gives pride of place to "blood" rather than common values or shared experiences in fundamental definitions of family.

The societal implications of this view can be seen, for example, in adoption law. In the past, American adoption practices were based on the idea that families were social units. Today, as genetic relationships are granted growing importance, adoption practices are changing accordingly. Guided by psychiatry, family court judges used to consider emotional ties as the most important factor in child custody decisions. Today, the emphasis has shifted to biological connectedness.

Genes as Determinants of Behavior

The second belief—that genes determine behavior—is reflected in notions about crime and punishment. Stories about "bad seeds"—ubiquitous in the eugenics-minded 1920s—persist today. In popular culture, personality traits and human behaviors are often attributed as much or more to "nature" than "nurture." Although no genes have been identified that affect traits such as success, kindness, criminality, or violence, "genes" in the abstract are an appealing way for people to explain individual differences, and race and gender stereotypes, as well as behavioral problems.

One area deeply affected by the belief that genes determine behavior is criminal law. If criminal behavior is seen as biologically-based, then people who are genetically predisposed to criminality cannot help themselves. In turn, society cannot hold them responsible for their criminal acts. Increasingly, popularized "biological" defenses—such as the influence of sugar or addictive drugs, premenstrual syndrome, or having an extra Y chromosome—are being used as strategies to gain acquittals or mitigate punishment in courts of law.

Predicting Future Health and Behavior

The third belief of genetic essentialism is that future behavior—and future health—can be predicted through genetic analysis. This belief legitimates efforts by schools, employers, and insurers—and many other groups—to anticipate and mitigate risk through identifying the predisposed.

At present, practical scientific knowledge about genes that predispose individuals to a particular disease or type of behavior is meager. It is clear that a handful of relatively simple genetic diseases—like Tay Sachs and Huntington's disease—are caused by single genes and tend to appear at a particular age and progress in a stereotypical way. However, most other noninfectious illnesses—like cancer, stroke, and heart disease—involve complex but poorly understood interactions between genetic and environmental factors. Moreover, no genes are known to predispose individuals to crime or violent behavior.

Nevertheless, the flow of action is fast outpacing the stock of fact. Some employers are attempting to use genetic rationales to make hiring decisions, and the first cases of "genetic discrimination" are being tried in the courts. Whether an insurer can deny medical coverage to people whose genomes include DNA-coding sequences that may predispose them to illness is being hotly debated. The controversies are likely to intensify in the coming years.

The Popular Appeal of Genetic Explanations

Why does opinion seem to diverge so far from fact? In large part, because genetic explanations are simple. They provide an unambiguous and appealing way to deal with questions of guilt and responsibility. Social problems can be laid at the door of an abstract entity—"the genes"—instead of requiring a more complex assessment of the social system, the family, or the individual. Or problems can be blamed on a particular demographic subgroup that is assumed to carry a "bad" set of genes.

At least twice in this century, genetic explanations have been used to jus-

> DNA is data without dimension, text without context, so that its equation with personhood requires a profound leap of faith. But DNA narratives appeal because they codify ambiguities about identity. They validate the individual as placed in unambiguous relationships—not of choice, but of blood.
>
> **Dorothy Nelkin**
> **New York University**

tify the tragic policies and ruthless operations of totalitarian regimes. The Nazis used genetics to enforce their claim to be the "master race" and justify the murder of "lesser" genetic beings. The Communists of the Soviet Union and China used Lamarckian genetics to support the idea that the Marxist utopia would produce a "new man" who would behave in ways acceptable to the Communist state.

Thus, it is crucial that both science and society pay close and vigilant attention to the ways in which scientific (or pseudoscientific) ideas influence our political, legal, and social thinking. ■

Commercialization of Biotechnology Innovations

T he pathway from a basic biological research discovery to a well-developed commercial product is long and complex, involving a host of scientific, economic, and legal obstacles. Many companies have lost considerable sums before getting a product to market. Many others have gone bankrupt. For the industry as a whole, however, the financial rewards continue to outweigh the risks. Indeed, the rate at which biotechnology innovations are transformed into products is greater than ever before.

Companies and Universities Need Each Other

Economic necessity is bringing industrial laboratories and university laboratories together. In a time of heightened domestic and global competitiveness, many companies are searching for less expensive ways to perform innovative research and create new technologies. Increasingly, those companies turn to academic laboratories to fill the "innovation gap." At the same time, the squeeze on federal funding is forcing universities to turn to the private sector for basic research funds. Alliances between companies and universities provide good opportunities for developing scientific discoveries into products that have a near-term impact on society.

Technology Transfer

The Bayh–Dole Act of 1980 gave U.S. universities the right to own inventions that arise from federally funded research projects. In most cases, the university licenses the invention to a company that develops the product further and brings it to market. Very often the universities and biotechnology companies remain in contact throughout the development of the product. In the eighteen years since the Bayh-Dole Act was passed, university/industry relationships have grown in number, size, complexity, and economic impact.

The University Investigator's Point of View

Some biotechnology innovations follow serendipitously from discoveries made in the course of studying basic biological phenomena, while others come directly from applied research projects sponsored by companies. From the university investigator's point of view, the process of moving the innovation out of the laboratory and into the marketplace is different in these two situations.

If the initial research was sponsored by a company, the investi-

---

Economic Forces Affecting Biotechnology Research During the 1990s

- The realization that university laboratories can no longer rely on the federal government as the primary source of research funding.

- Consolidation and down-sizing of the pharmaceutical industry, due to:
  > Constraints imposed by managed care providers.
  > Uncertainties concerning proposed federal health care bills.

- Decreased investment in biotechnology companies by private investors.

- Increased investment in biotechnology by pharmaceutical companies (about $2.2 billion in 1995).

> From an investigator's point of view, the trickiest part is deciding whether a discovery is commercially important. It depends on whether we can anticipate the ultimate value of a discovery in a science in which our understanding of how things work is constantly changing. In plant biotechnology, we have wasted some opportunities.
>
> **Nam-Hai Chua**
> **Andrew W. Mellon Professor**
> **The Rockefeller University**

gator informs the sponsor and the university's technology transfer office. The investigator then writes a report describing the invention before filing a patent. In most cases, the legal costs of writing, reviewing, and filing the patent is borne by the sponsor. The university owns the patent, and the sponsor has the first right to license it.

If the initial research was not sponsored by a company, the investigator and the technology transfer office must decide if the discovery is important enough to patent. If they decide to proceed, the technology office will help identify interested companies who may bear the costs of the patent in exchange for exclusive licensing rights.

The ongoing relationship between the university investigator and the industrial sponsor is shaped by the agreements they negotiate when licenses are granted. The investigator may ask for:

- Payment of the patent application fee and patent maintenance fees.
- An upfront payment in the form of a technology transfer fee.
- A research grant to support further work on the product.
- A good faith commitment to commercialize the technology.
- A royalty stream to the university.

If further work is required to develop the invention into a product, the investigator and sponsor may agree to support further development work in the investigator's laboratory; transfer the innovation to the sponsor's laboratory; or continue to collaborate, with the investigator acting as consultant to the company.

The Company's Point of View

While most researchers in the biotechnology industry are trained in university laboratories and are familiar with academic expectations, some basic researchers do not understand the needs and perspectives of industry scientists. Nevertheless, it is increasingly true that small biotechnology companies, large pharmaceutical companies, and academic laboratories are coming to appreciate—and depend upon—one another's strengths.

> Intellectual property protection is not just to fill the pockets of biotechnology executives, but also to allow scientists to see their knowledge converted into applications.
> **Joan S. Brugge**
> **Senior Vice President and Scientific Director**
> **Ariad Pharmaceuticals, Inc.**

From the company's point of view, cooperation between academic and industrial laboratories is enhanced when:

- University technology transfer offices educate their faculties to be aware of the potential commercial applications of their research.
- University technology transfer offices are familiar with biotechnology and pharmaceutical companies and can match up laboratories and companies.
- Academic and industrial laboratories work together closely to translate basic science innovations into products.
- University scientists keep patentable discoveries confidential, even when they write grant applications or speak at scientific meetings.
- Licensing agreements are tailored to fit the specifics of the collaborative situation.
- University researchers avoid conflicts of interest, especially if they hold stock in biotechnology companies.

> The classic investment model of venture capital—in which small groups of investors fund companies in many small rounds of capital—no longer works for the biotechnology sector. Investors are coming together in creative ways to form larger syndicates (often with large corporations) to better share risks and to provide adequate financing.
>
> **Sandra Panem**
> **President**
> **Vector Fund Management**

The Investor's Point of View

Investors balance the potential rewards of a given investment with its apparent risk. They choose between high-return, high-risk investments (like new biotechnology ventures) and low-return, low-risk investments (like treasury bonds). According to investment advisors, a high-risk venture by a start-up company should promise a potential return of 35–40 percent. In comparison, a new venture by an established company should promise a return of about 25 percent, while an investment in the stock market should promise a return of about 8–15 percent.

Although the biotechnology industry is barely twenty years old, it already has a complex history and is starting to "mature" as an investment market. The first biotechnology companies appeared in the mid-1970s, and many new ones appeared throughout the 1980s. There was great excitement and optimism, leading to over-expansion and the overvaluation of many companies. In the 1990s reality set in, and a consolidation (or "rationalization") of the industry followed. Many small companies went bankrupt or merged into larger ones.

By 1996, the biotech sector included about 294 public and 1,287 private companies. Between 1985 and 1996, biotechnology sales rose from $1.1 billion to $10.8 billion, and revenues increased from $2.2 billion to $14.6 billion (see *Biotech 97: Alignment*. Ernst and Young, 1995). Moreover, biotechnology continues to be a highly volatile sector in which large amounts of money can be both made and lost. While the outlook for the industry as a whole looks bright, individual investment can be extremely risky. Collectively, biotechnology companies lost $4.1 billion in 1997.

The biotechnology sector is still primarily composed of many small entrepreneurs and continues to be an exciting source of high-risk/high-return investments. Despite past setbacks and continued volatility, investors view biotechnology as a "sustainable" industry, and an area in which the United States is likely to remain the world leader for some time to come.

Moreover, with time has come wisdom. Investors now understand the health biotechnology industry better than they did in the 1980s. They know that success is dependent on years of basic research and that many drugs require at least eight years of development and testing before coming to market. Today, health biotechnology attracts well-informed, long-term investors.

Public Policy Issues Affecting the Biotechnology Industry

Washington's changing political climate may influence many of the financial and social parameters affecting the biotechnology industry. However, because the industry has already overcome many obstacles,

---

Health Biotechnology Investment in the 1990s

Although investment advisors believe it is a good time to invest in health biotechnology ventures, they recommend that their clients be very selective, especially when providing capital at the earliest stages of a new venture. Professional investors recognize that there is not enough money to develop every innovation. Therefore, they may review as many as 300 potential ventures before choosing one to support. This statistic underscores how difficult it is for a small entrepreneur to attract investment capital for biotech ventures in the 1990s.

---

few of these factors are likely to have a major impact in the near term. Longer-term trends are harder to predict.

Changes in Tax Policy

Several business-friendly proposals may, if sustained, improve the availability of capital for investment in biotechnology. These include the R&D tax credit, a capital gains tax cut, and perhaps an orphan drug credit that can be carried forward in successive fiscal years.

Federal Funding for Basic Research

Strong federal support for biomedical research has been one of the underpinnings of U.S. preeminence in commercial biotechnology, and continues to be important to the overall long-term health of the biotechnology industry. Political support for the National Institutes of Health (NIH) has survived periodic budgetary attack, and the NIH was the only federal S&T agency to receive a major increase in its budget for FY1997. Congressional and public support for basic biomedical research remains strong and broad based. Nevertheless, the days in which a substantial proportion of academic scientists could depend on continued government support—as long as they remained productive and respected by their peers—are long over.

Technology Policy

Federal support for technology programs at the NIH and the National Institute of Standards and Technology (NIST)—including the Small Business Innovation Research Program, the Advanced Technology Program (ATP), and the Manufacturing Extensions Partnerships (MEP)—have come under fire since the Republican takeover of the Congress in 1994. The purpose of ATP is to enable private industry to undertake cost-shared pro-

jects that are risky, but have
potentially important long-term
economic benefits. MEP provides small manufacturers with
access to new technologies.
These programs are young and
their impact is hard to measure.
While they may survive, their
future growth and role in development of the biotechnology
industry is uncertain.

> There is recognition throughout
> the government that the transfer
> of basic research to industry is a
> positive goal that should not be
> encumbered by burdensome,
> bureaucratic approvals or con-
> tingency clauses.
> **Roger C. Herdman**
> **Former Director**
> **Office of Technology Assessment**

Technology Transfer Policy

The controversy over property and development rights to inventions generated by government-funded laboratories has largely been
resolved. Today, university researchers and their institutions routinely license innovations to companies for commercialization.
Federal government support for biomedical technology transfer is
strong and largely bipartisan.

Food & Drug Administration (FDA) Reform

A series of proposed changes in FDA rules and procedures is
likely to have a positive and near-term effect on the U.S. biotechnology industry. These changes include:

- Relaxing some regulatory controls in manufacturing.
- Relaxing some requirements for clinical trials.
- Implementing further strategies to decrease processing
  times for new drug approvals.
- Permitting the collection and submission of confirmatory
  data after a drug has been marketed.

The FDA is not likely, however, to accept several other proposed
changes that it believes will undermine its ability to protect the pub-

lic. These proposals include accepting European drug approvals without further examination, employing non-FDA reviewers and appeal boards, and allowing the submission of summary data on a new drug instead of a complete dossier of raw data.

Public Perception

Surveys indicate that the U.S. public overwhelmingly supports biotechnologically derived pharmaceuticals, gene therapy, and, despite initial skepticism, bioengineered foods. If they are sustained, these views will provide a boost to the industry and to the investment community.

Intellectual Property Protection

Since the National Bioethics Advisory Committee was established last year at the urging of Senator Mark Hatfield of Oregon, earlier concerns about patenting engineered life forms seem to have subsided. No major statutory changes that affect biotechnology patents are expected in the near future, beyond those required to bring the United States in line with provisions in the General Agreement on Tariffs and Trade (GATT), which has metamorphosed into the World Trade Organization.

Biotechnology and Patent Utility Guidelines

U.S. patent law requires that a discovery must be "useful" to be patentable. During the nineteenth century (before modern regulatory agencies were established), this provision, known as the "utility requirement," was used to protect the public from detrimental inventions.

The modern understanding of the utility requirement derives from the 1966 Supreme Court case, *Brenner* vs. *Manson,* which concerned a patent application for a process for manufacturing a steroid whose medical utility was inferred from its structure but which had not been tested *in vivo*. In rejecting Manson's appeal, Justice Abraham Fortas ruled that to be useful the invention must have a demonstrable "end-use." That is, it must provide a "specific benefit" to consumers beyond its benefit to researchers.

This ruling poses particular problems for the health biotechnology industry. Because many biotechnology inventions are intended to treat conditions that have never been treated before, there may not be any well-established assays in animal models or *in vitro* that have a proven track record for indicating the therapeutic effect in humans. Thus, if the claimed utility is human therapy, the Patent and Trademark Office (PTO) requires applicants to demonstrate the product's end-use by conducting clinical trials. This requirement has put small biotechnology companies in a double-bind: A company may need investment capital to conduct a clinical trial so it can apply for a patent, but it cannot interest an investor until it has acquired one.

At the urging of the biotechnology community, the PTO reviewed this issue, and new guidelines for examining patent utility went into effect in July 1995. The guidelines provide that "the examiner should accept any reasonable use that can be viewed as providing a public benefit." The guidelines also hold that a reasonable correlation between a test for pharmacological activity and therapeutic benefit is sufficient, and that human clinical trials are not required

New Guidelines Affect Research Uses of Patented Inventions

Assuming that the new guidelines are upheld by the Supreme Court, health biotechnology innovations will more easily meet the utility requirement in patent law. However, the new guidelines may have unintended negative effects in the basic research community.

> **Allowing an experimental use defense for academic researchers is an idea that would surely be considered seriously if it were made by the very same people who argued so successfully for the new utility guidelines.**
> *Rochelle C. Dreyfuss*
> **Professor of Law**
> **New York University Law School**

By rejecting a contribution to research as a "utility," the *Brenner* vs. *Manson* ruling had the effect of leaving many tools of research in the public domain. However, the new guidelines imply that academic and as well as industrial researchers are required to license each patented product that they use. The costs and paperwork involved in complying with these guidelines could place a new financial burden on universities.

There are ways to avoid this problem. Many biotechnology products have both a research use and an end-use. Ideally, an end-use should be subject to a patent while a research use remains in the public domain. This could be achieved through a statutory change in patent law. Alternatively, the courts could permit university researchers to employ an "experimental use defense" against accusations of patent infringement.

Conclusion

As the preceding discussion has illustrated, shepherding a biotechnology innovation from the laboratory to the market involves the concerted efforts of numerous individuals from a variety of backgrounds—including university scientists, business executives, technology transfer "matchmakers," industrial scientists and engineers, patent attorneys, and investors. Lured by the double promise of scientific excitement and financial success, these hardworking individuals come together to embark on the long and complex process of transforming a biological discovery into a commercial product.∎

Biotechnology in New York

During the 1980s, numerous small biotechnology companies grew up near universities and academic medical centers in large cities such as Boston and San Francisco. Despite its deep reservoir of life sciences resources, however, New York City was not among the emerging biotech centers. This chapter discusses why New York has lagged behind and suggests what might be done to make New York a more hospitable environment for the biotechnology industry.

The first section—Biotechnology and Public Health in New York City—provides past examples of public–private collaboration to stimulate the development of medical technology. It also describes how the city's public health needs have historically influenced the direction of biomedical research. The second section—Biotechnology and the Economy of New York—considers the potential economic benefits of an expanded biotechnology industry and reflects on what went wrong in New York. The chapter concludes with a set of recommendations gleaned from the discussions at the colloquia and from other studies, including some sponsored by the New York Academy of Sciences.

Biotechnology and Public Health in New York City

Although New York City has serious problems that hinder the

regional development of the biotechnology industry—including
the lack of affordable laboratory space and a host of quality-of-life
issues—the city possesses the largest concentration of first-class
medical and research institutions in the world. Moreover, the social
and public health problems that continually challenge the metropolitan area have also made New York City a hands-on laboratory
for the study of national urban issues and the development of solutions. In fact, the public health problems of the city have been—
and continue to be—drivers of innovation in biomedical research
and biotechnology. Many recent public health initiatives have been
based on innovative biotechnology products, including new antibiotics and vaccines, new and improved contraceptive tools, and
rapid diagnostic tests for communicable diseases.

Public–Private Collaboration: Two Historical Precedents
<u>1941: The Public Health Research Institute</u>. New York City has
a long tradition of partnership between the public and private sectors to promote technological advances to address public health
problems. In 1941, Mayor Fiorello La Guardia founded the Public
Health Research Institute (PHRI), the first urban public health
research institute in the United States. At the urging of
Commissioner of Health John L. Rice, the city provided an annual
payment of $100,000 per year for ten years, as well as free housing,
laboratory space, and office facilities.

By the 1950s, PHRI's annual appropriation from the city had
quadrupled to $400,000 plus housing and maintenance services.
The Institute had a staff of 90 people working in five divisions:
infectious diseases, applied immunology, nutrition and physiology,
epidemiology, and laboratory diagnosis. Among the many significant biotechnological contributions of PHRI was the discovery by
Jules Freund that Jonas Salk's newly developed polio vaccine was

more effective when used with a water-in-oil adjuvant.

1958: The Health Research Council. The city established the Health Research Council (HRC) in 1958 to support research in universities, colleges, medical schools, and city agencies. HRC's goal was to pioneer new ways to solve public health problems related to childbirth and infant care, gerontology, mental illness, accidents, heart disease, chronic pulmonary

> We appear to be entering a period of great instability, in which the kind of public courage that saw the establishment of PHRI and HRC has eroded, social support for the long-term benefits of scientific research grows increasingly problematic, and funding streams continue to shrink.
> *Margaret A. Hamburg*
> **Commissioner of Health**
> **New York City**

disorders, environmental hazards, and drug addiction. HRC also aimed to support young research scientists and to promote scientific education and training.

HRC was created by Mayor Robert F. Wagner and Commissioner of Health Leona Baumgartner, in collaboration with the National Institutes of Health, the Rockefeller Institute for Medical Research (now The Rockefeller University), the Rockefeller Foundation, several medical schools, and private philanthropists. The city allocated $600,000 to the first year's operation, with the hope that the budget would eventually increase to one dollar per citizen per year, or about $8 million a year.

In 1962 HRC supported the broad-scale testing of a measles vaccine and expedited its licensing and approval. Through the 1960s, HRC supported research in nutrition, mental illness, chronic diseases, sickle cell anemia, pollution from asbestos, and the biology of drug addiction. Moreover, it was through the support of HRC that the usefulness of methadone was discovered, resulting in an enormous advance for the treatment of heroin abuse.

Implications for Today

PHRI and HRC were established by people of vision who saw opportunities as well as challenges and created structures to take advantage of both. Unfortunately, budget pressures threatened both institutions and led to the demise of HRC.

Today's circumstances offer limited evidence on which to base renewed optimism. If the New York metropolitan area is to remain one of our nation's most influential centers of research and public health leadership, an extraordinary level of cooperation and a fresh commitment will be required among the scientific, medical, and public health communities, and between the public and private sectors.

Biotechnology and the Economy of New York State

Of an estimated 1,300 biotechnology companies in the United States, only 140 are located in New York State. While the growth rate for biotechnology companies has picked up in recent years, many significant challenges face the biotechnology industry in New York.

Why Should New York Care?

Biotechnology is a growing industry that could bring great economic benefits to New York City and New York State, as it has to other regions.

R&D Investment. Nationally, the amount of investment in research and development by biotechnology companies is substantial, accounting for nearly 25 percent of the total of $30–31 billion in R&D expended in 1994. It is estimated that biotechnology

---

Biotechnology Industry Creates Jobs

Molecular biology is an extremely labor-intensive type of scientific work. Because the biotechnology industry is based on molecular biology R&D, it is a labor-intensive industry.

According to the U.S. Department of Commerce, seven of the ten U.S. companies that spent the most money on R&D per employee in 1994 were biotechnology companies. In fact, on average, biotechnology companies spend 49 percent of their revenues on R&D—more than any other industry, including the pharmaceutical industry.

According to a study by the Bank of Boston, each million dollars invested in biotechnology creates 25.5 jobs. In this category—jobs created per million dollars—biotechnology ranks 48th out of 333 manufacturing industries.

---

companies will spend more than $50 billion in R&D by the year 2000. These private sector investments can help to offset decreases in federal funding of academic research.

Employment. The biotechnology industry may provide society with the largest number of high-skilled, high-wage jobs of any growth industry over the next ten years (see box above). Although the biotechnology industry in New York State is presently relatively small, accounting for only about 30,000 jobs, it employs a high percentage of skilled staff, including scientists, lawyers, and business executives.

Industry–University Collaboration. When inventions developed in academic laboratories are patented and licensed to biotechnology or pharmaceutical companies, the economic benefits can be impressive. Licensing agreements between academic centers and companies in the United States currently generate about $20 billion per year in R&D investments, creating an estimated 160,000 jobs. Thus, even though it takes 8 years on average to develop a new biotechnology product, licenses stimulate investment and promote jobs even *before* a product reaches the market.

Economic Growth Potential. The biotechnology industry

includes a wide range of applications and products and has enormous potential for expansion. In some industrial sectors—such as agriculture—biotechnology products are just starting to come to market. This means that there may still be time for New York to "catch up" and develop a significant biotechnology sector.

In 1997, the total worldwide sales of biotechnology products totaled $13 billion, including $7.7 billion within the United States. The major components of worldwide sales include:

- $5.6 billion in drugs for the treatment of human diseases.
- $1.5 billion in human diagnostics.
- $4.9 billion in biochemicals.

As of late 1997, 65 drugs and vaccines have been approved by the U.S. Food and Drug Administration; more than 295 are currently in human clinical trials; and hundreds more are in early development in the U.S. In addition, the FDA has approved over 1,300 diagnostic products, including home pregnancy and ovulation tests.

In future years, environment-friendly industrial and agricultural products will add greatly to this profile. Moreover, biochemical products generated by novel biotechnological manufacturing processes are expected to be especially profitable. Biochemicals include vitamins, amino acids, enzymes for food processing, and biodegradable polymers.

What Went Wrong in New York?

The problems of small entrepreneurs in New York are not necessarily specific to biotechnology enterprises. Space limitations, rental costs, transportation problems, public education

> In New York, we missed an absolutely critical decade of industry growth and company formation.
> *Jack Huttner*
> Executive Director
> New York Biotechnology Association

system weaknesses, and general quality of life concerns, were all cited regularly as impediments to a thriving high-tech entrepreneurial sector. More recently, while some trends suggest improvement in the business environment, fostering economic growth has become even more an imperative as New York struggles to emerge from almost a decade of recession. For biotech entrepreneurs, four additional problems compound the problem of start-up ventures.

Inertia and Indifference. Fostering entrepreneurship has not been a political or economic priority. The economic development focus in New York has been on very large companies. The importance of biotechnology to future global competitiveness, and the essential role of small, high-tech companies in building a biotechnology industry, has been little recognized.

A Declining Share of NIH Funds. New York State's share of NIH research funds has declined over the last two decades. In 1997, New York institutions received 10.7% of the $9 billion made available by the NIH for research, down sharply from the 15.2% New York was awarded in 1983. In contrast, Massachusetts and California both saw their share of NIH funding remain fairly steady at around 11% and 15% respectively.

The overall decline in research funds has led to difficulties in supporting basic research and recruiting young scientists to New

---

Industrial Links and New York City's Universities

According to the Association of University Technology Managers, New York City's research universities are lagging behind comparable institutions in the United States in their funding from industry. For example, in 1993, New York University Medical Center ranked 60th in industry-sponsored research expenditures ($7 million), while Columbia University ranked 67th ($6 million). These figures included clinical trials of pharmaceutical products.

> While New York City once was the most heavily funded city in the nation, the growth of its research funding has not kept pace with incremental growth in the NIH's budget. Further, its growth rate in obtaining federal money is the lowest of all top NIH-funded cities.
> *Report of the Commission on Biomedical Research and Development, 1993*

York City medical centers. Tenured faculty have been forced to curtail their research activities and many junior and mid-level faculty have opted out of research entirely. Without a thriving basic research base, a regional biotechnology industry cannot prosper.

"Cultural Impediments" in Academia. Many observers have remarked upon the past failure of New York's academic centers to build strong relationships with private companies. The lack of a deep corporate–academic partnership history has impeded biotech development in the city. With financial pressures becoming ever more acute, New York City's biomedical institutions are now beginning to encourage technology transfer relationships between their faculty members and industry scientists.

Threats to Urban Academic Medical Centers. The restructuring of the health sector and the advent of managed care have altered the traditional stream of revenue that has supported New York's academic medical centers.

As tertiary care centers that employ a high percentage of medical specialists and serve as teaching hospitals for interns and residents, these centers tend to have higher unit costs than other facilities that compete for managed care contracts. In addition, managed care rates do not factor research and teaching functions into reimbursement structures. Hence, the clinical research centers which train biologists and foster medical innovations are in deep financial crisis, as is the financing of their research and teaching functions.

The academic medical research centers in New York City not only face operating budget constraints, but are also starved for capital to maintain and improve their plants and equipment. Around the country, academic health centers have invested substantial funds in their research laboratories and have been better prepared than New York City's institutions to form partnerships with companies. Moreover, some have had less rigidly controlled health care reimbursement arrangements. Thus, medical centers at universities elsewhere may be finding organizational patterns better suited to the nation's future health system.

> Academic health centers in the New York region are functioning under enormous pressure, with all their funding streams under threat. A decided and committed group of people from multiple entities in New York could make a big difference by bringing society that message.
> *Herbert Pardes*
> **Vice President & Dean**
> **Columbia University**

Recommendations for New York

Nearly everyone agrees that an infusion of high-tech biotechnology companies would bring much-needed jobs and investment to New York State and New York City. Attracting these businesses would require changes in New York's political, academic, financial, and

---

The Audubon Center: The First Biotechnology Incubator in New York City

The State University of New York at Stony Brook and the Rensselaer Polytechnic Institute in Rochester, New York, have operated successful biotechnology incubators for a number of years. However, until very recently, there were no incubators in the New York metropolitan area.

After overcoming financial and political obstacles (which took many years), New York State and New York City have joined with Columbia University to open the Audubon Center in upper Manhattan. As of mid-1998, fourteen companies, twelve of which are biotechnology firms, are located in in the Audubon Center.

---

business communities. Ideally, a diverse group of individuals and institutions would join the effort, including city and state politicians; venture capitalists; local medical schools and universities; small biotechnology companies; and large pharmaceutical companies. The effort would require a partnership of vision, objectives, and resources over a sustained period of time.

The following action agenda for making New York a more hospitable environment for biotechnology emerged at the colloquium.

(i) State Policy Initiatives. Supportive state policy on capital formation, tax reform, incubation strategies, and economic development is an important foundation for a thriving biotechnology industry that increases New York's share of global trade.

(ii) Public Education. Biotechnology is a clean, simple, ecologically safe, and relatively cheap industry. It is also a good way to increase the number of high-paying, highly skilled jobs and to increase tax revenues for the city and state. Greater efforts are needed to educate the public, elected officials, and the business and financial communities about the potential of biotechnology in the city and state economy.

(iii) Improve technology transfer. Universities and medical centers need to make a concerted effort to encourage academic–industrial collaboration. For example, universities could encourage faculty members to participate in networking efforts to bring together industrial and academic scientists for the joint pursuit of NIH's Small Business Innovative Research (SBIR) grants.

(iv) Support Research. Private corporations and academic medical centers need to be more closely allied. Sharing equipment, fostering

exchanges of staff, and enhancing long-range joint R&D are among the steps needed to create a more vibrant research community.

(v) Encourage and support academic medical centers. Healthcare providers at the state level should work together to maintain quality, access to care, and a base of innovation. Pharmaceutical and biotechnology companies must join with academic medical centers and health care organizations to support and maintain the excellence of the research, teaching, and medical enterprise.

(vi) Support young investigators. A fund could be created to provide support to young biomedical research investigators so as to attract the next generation of researchers to New York City. This fund should be large enough to support a significant number of junior scientists and compensate for the loss of federal funding that has caused many able researchers to leave the field.

The New York Biotechnology Association

The New York Biotechnology Association (NYBA) is an industry-supported organization whose mission is to create conditions that will allow the biotechnology industry to prosper and grow in New York State. NYBA represents the interests of biotechnology companies, pharmaceutical companies, research universities, suppliers, and industry consultants.

NYBA aims to serve its constituencies by providing legislative representation and advocacy in Albany and New York City. It also provides economic development assistance and fosters technology transfer and commercialization opportunities among public and private sector groups. In addition, NYBA works to improve the public perception of New York State as a good place for small and expanding biotechnology companies. Finally, NYBA disseminates information and policy analysis on issues that are important to the biotechnology community.

The European Model for Technology Transfer

The major impediment to the growth of the biotechnology industry in Europe is the low level of venture capital, which makes it almost impossible for individual scientists to transfer inventions made in university laboratories to the commercial sector. In addition, the European patent laws hamper the development of new biotechnology projects, as the ownership of the patent usually resides with the inventor and is rarely transferred to the university. Thus, European academic centers have little incentive to help scientists commercialize products.

Because of these limitations, Europeans have mostly supported technology transfer through government channels. In France, "Biofuture," a $300 million, government-supported program, has brought together a government-funded pharmaceutical company (Rhone-Poulenc), the national research councils for science and medicine, and the Pasteur Institute. In Germany, the previous Ministry for Research and Technology invested billions of dollars in developing biotechnology at universities and small companies through specially supported activities. Finally, over the last several years, Sweden's Board for Technological Development, which has a yearly budget of $200 million, has supported the biotechnology effort with as much as $1 billion—a substantial amount for a small country with a population the size of New York City.

(vii) Encourage the establishment of biotechnology incubators. In Boston, Seattle and San Francisco—which provide significant assistance to fledgling enterprises—many small biotechnology companies have sprung up near large academic research institutions. In emulation of those cities, New York should work with local academic research centers to establish biotechnology incubators to foster the development of small, innovative companies. For example, the state or city might establish programs that provide start-up funds for new companies, or that support the renovation of laboratories in close proximity to universities.

(viii) Address regulatory barriers. Regulatory barriers provide significant disincentives to biotechnology incubators and small start-up companies. Some regulations also hamper the expansion of small

The Commission of Biomedical Research and Development

The Commission on Biomedical Research and Development was created by Governor Mario Cuomo in 1991 as a public–private sector initiative to enhance the competitive position of the academic and for-profit biomedical research community in the New York City region. In its report, the Commission prescribed immediate action to recapture New York's leadership position in biomedical research.

With the Commission's advocacy and support from State Senator Roy Goodman, the New York Biomedical Research Council was established in 1995 under the auspices of the New York Academy of Medicine. Its aim is to implement the Commission's twin goals of strengthening biomedical research and fostering the development of an expanded biotechnology industry.

companies that are beginning to grow and prosper. A careful inventory of regulations must be undertaken to ensure that the protection of public welfare inherent in regulatory intent is made compatible with industrial investment.

(vii) Encourage cooperation among academic health science centers in developing diagnostic tests and treatments for major human diseases. To solve complex (and expensive) research problems, universities within New York City might merge some of their technology offices or research units to improve interdisciplinary interaction in important areas. Alternatively, the universities might create a coordinating service to "match-up" laboratories that are working on different aspects of the same problem.

Actions such as these will require changes in attitude as well as administration. Instead of competing against one another for scarce resources, academic laboratories in New York could pool their resources and work together to re-establish New York as a leading center for clinical research.

(viii) Create a city or state Office of Technology Assessment (OTA).
A New York Office of Technology Assessment could evaluate the economic competitiveness of the biotechnology industry in the metropolitan region and identify additional strategies for strengthening New York's commitment to biomedical research and technology. It could help journalists, politicians, city planners, health care organizations, and venture capitalists assess the economic implications of the major scientific breakthroughs that occur over the next decade.

The New York Academy of Sciences may be able to play a significant role in pursuing these opportunities. In the last two years, the Academy has initiated a series of efforts to deepen the linkage between science and technology and economic development in the New York region.

A November 1995 regional conference, co-convened by the Academy with the Federal Reserve Bank of New York—and with the collaboration of the Regional Plan Association, the Port Authority, the Bureau of Labor Statistics for the New York Region, and the Partnership of the City of New York—explored prospects for regional economic development based on encouraging and exploiting technological innovations. With support from the Lounsbery Foundation and the Alfred P. Sloan Foundation, the Academy is following up on that conference with a series of discussion seminars in 1996 and 1997 focused on critical issues associated with the development of business-university-government partnerships and with understanding the impact of technology-based economic development strategies on the region's employment base. ∎